"SREE UDHYAN WRITINGS"

The Cosmic Chronology

"Space and Time: Everything from the beginning of the universe"

SREE UDHYAN

Author: Sree udhyan raam M

This book is totally based on research and does not violate any of the copyright issues. The facts and the matter are based on true research and are well revised. The author and the book are not responsible if any matter matches with some other source, and it would be a pure coincidence. Any content related to this book in future will be punished

©All rights reserved.

Copyright owner

Maddineni Sree Udhyan raam

FROM THE CREATOR OF:

LIMTER: THE SUPERIOR LEGEND

ISBN: 9798341093409

Contents

- Introduction to astronomy
- Beginning with astronomy
- What is Astrophysics?
- What is cosmology?
- Modern instrument we use
- About the universe
- The traces of the beginning -1 : The Big Bang
- The traces of the beginning -2 : Forming stars and galaxies in the young universe
- Classifying stars with spectral analysis
- Over to the end of the stars
- Discovering the Undiscovered: Blackholes
- Galactic odyssey : Exploring the Cosmos
- Trails of the Milky Way
- The formation of the Solar System

- The Messenger Planet: Mercury
- The Evening Star: Venus
- The Planet of Wonders: Earth
- The Red Neighbour: Mars
- The Dictator and the Saviour: Jupiter
- The Cosmic Elegance: Saturn
- The Blue Giant : Uranus
- The Last Knight : Neptune
- Guardians and Separators: Asteroid and Kuiper Belts and Beyond
- Dwarfs of the Solar System
- An Unexpected Miracle
- The Early Conquerors: The age of Dinosaurs
- Advancement in Genes
- Over the Time of Humans in Astronomy
- Kepler's Laws of Planetary Motion
- Edwin Hubble and Doppler Swift

- Unveiling the Big Bang: A Tapestry of Evidence
- The Frontier of the Cosmos
- Satellites: Natural and Artificial
- Astrobiology: Extraterrestrial Life
- "Colonization in the Future!"
- Exploring the Universe
- The Last of Celestials: Nebulae
- Stars, Red Hyper Giants, Brown Dwarfs and Sub-Stellar Objects
- Binary Star System
- General Theory of Relativity {Corroborated}
- Quasars and the Older Light
- Scaling the Observable Universe
- Tribute to Worlds Geniuses and Machines
- Top Space Research Organizations in the world
- The End of the Universe

Author's note

"We have done a lot of research on revising topics and theories. Please pay enough attention towards your passion and your dreams. Follow every sentence of the book. Remember, success never walks to you, you must make a way"

-Sree Udhyan Ram .M

Astronomer and author

Age: 14(2024)

"I would like to dedicate this work to the people who supported me, who suggested me, who led me, who saved me, and who accompanied me. First in my list comes my father, Maddineni seetharam, a software engineer, who had supported me and led me into all new and different life style. He gave me a completely new way to achieve in my life. He guided me into a new way and introduced me to another shade in me; he was the responsible person for me today being an author. Then, my queen in my life, my mom, Maddineni Swpna, who took me on to this tough world and taught me the life teachings that she experienced. She had allowed me to be an author but with no differences, my brother who had been the cutest thing in my life till date, he found me in emotions with the moment that had happened together, he is the person to be the one till my end as my partner. My teachers supported me as well as gave me the power to know and understand things. My friends who suggested me, who helped me, and who encouraged me, of them made me experience extreme friendship but some took a heart place. I found this fantastic when these many people supported me. I wish I would be lucky enough to have these people around me until I leave this world. Once again, I thank all the people for your support for my success."

10 | THE COSMIC CHRONOLOGY

PRIOR NOTE : DISCLAIMER

The information in this book is based on facts researched from a variety of internet sources. Copyright applies to creative expression, and facts are not protected by copyright. Every effort has been made to ensure the accuracy of the information presented.

12 | THE COSMIC CHRONOLOGY

Chapter 1
Introduction to astronomy

Introduction to Astronomy:

Astronomy, the study of celestial objects and phenomena beyond Earth's atmosphere, has captivated human curiosity for millennia. From ancient civilizations gazing at the night sky to modern space exploration, the quest to understand the cosmos continues to inspire wonder and awe. At its core, astronomy seeks to unravel the mysteries of the universe, exploring everything from distant galaxies to the fundamental forces shaping our cosmos. Through the lens of astronomy, we gain insight into the origins of the universe, the life cycles of stars, the formation of planets, and the vastness of space itself. In this introductory exploration, we embark on a journey through the cosmos, delving into the tools, theories, and discoveries that define the field of astronomy. From the familiar constellations overhead to the cutting-edge research probing the depths of space, let us embark on a voyage of discovery and enlightenment, guided by the timeless wonders of the universe. Astronomy, the oldest of the natural sciences, has captivated humanity for millennia with its exploration of the cosmos. From ancient civilizations gazing at the stars to modern-day space missions probing the depths of the universe, astronomy encompasses the study of celestial objects and phenomena beyond Earth's atmosphere. At its core, astronomy seeks to understand the origins, evolution, and fundamental workings of the universe, from the smallest particles to the largest galaxies. It encompasses a vast array of topics, including the motions of planets, the life cycles of stars, the formation of galaxies, and the mysteries of dark matter and dark energy. By observing and analyzing the cosmos through telescopes, satellites, and space probes, astronomers unlock the secrets of the universe, expanding our knowledge and reshaping our understanding of our place in the cosmos. Whether peering through a backyard telescope or scrutinizing data from cutting-edge observatories, the pursuit of astronomy continues to inspire wonder, curiosity, and exploration across cultures and generations.

15 | THE COSMIC CHRONOLOGY

Chapter 2
Beginning with astronomy

At the dawn of human civilization, our ancestors looked up at the night sky with wonder and awe. In those early days, astronomy was born out of necessity as well as curiosity. The movements of the sun, moon, and stars provided vital markers for agricultural and navigational purposes, guiding the rhythms of daily life and the cycles of planting and harvesting. Ancient cultures around the world developed elaborate systems of astronomy, weaving celestial observations into their mythologies, calendars, and religious practices. From Stonehenge in England to the pyramids of Egypt, ancient monuments bear witness to humanity's ancient fascination with the heavens.

As civilizations evolved and knowledge expanded, so too did our understanding of the cosmos. Early astronomers such as Aristotle and Ptolemy developed theories to explain the motions of celestial bodies, laying the groundwork for the scientific revolution that would follow centuries later. With the invention of the telescope in the 17th century, astronomers gained unprecedented access to the heavens, revealing a universe far more vast and complex than previously imagined. Galileo's observations of Jupiter's moons and the phases of Venus challenged long-held beliefs about Earth's central position in the cosmos, sparking a paradigm shift in our understanding of the universe.

Since then, astronomy has continued to advance at an astonishing pace, fueled by technological innovations and international collaboration. From the discovery of distant galaxies and black holes to the exploration of exoplanets and the cosmic microwave background radiation, each new breakthrough expands our horizons and deepens our appreciation for the beauty and complexity of the cosmos. Today, astronomers study the universe across the entire electromagnetic spectrum, from radio waves to gamma rays, using ground-based observatories, space telescopes, and sophisticated instruments to probe the mysteries of the universe.

As we stand on the threshold of a new era of discovery, the study of astronomy continues to inspire wonder and curiosity, reminding us of our place in the vast tapestry of the cosmos. From the birth of stars to the evolution of galaxies, astronomy offers a window into the history of the universe and the forces that shape our cosmic destiny. As we gaze up at the night sky, we are not just observers but participants in a grand cosmic drama, connected to the stars by the same elemental forces that govern the universe. In the words of Carl Sagan, "The cosmos is within us, we are made of star-stuff, and we are a way for the universe to know itself."

19 | THE COSMIC CHRONOLOGY

Chapter 3
What is Astrophysics?

Astrophysics is a specialized field of study within astronomy that focuses on understanding the physical properties and behavior of celestial objects and phenomena using the principles of physics. It seeks to unravel the mysteries of the cosmos by applying the laws of physics to astronomical observations and theoretical models. Unlike traditional astronomy, which primarily involves observing and cataloging celestial objects, astrophysics delves deeper into the underlying physical processes that govern the universe.

At its core, astrophysics encompasses a wide range of topics, including the formation and evolution of stars, galaxies, and planetary systems, the nature of dark matter and dark energy, the structure and dynamics of the universe, and the study of exotic objects such as black holes, neutron stars, and supernovae. By investigating these phenomena, astrophysicists aim to answer fundamental questions about the origin, structure, and fate of the universe. Another key area of study in astrophysics is cosmology, which seeks to understand the large-scale structure and evolution of the universe as a whole. Cosmologists investigate questions related to the origin and evolution of the universe, the nature of dark matter and dark energy, the cosmic microwave background radiation, and the ultimate fate of the cosmos. By analyzing the distribution of galaxies, measuring their red shifts, and studying the cosmic microwave background, cosmologists can piece together a detailed picture of the history and composition of the universe. In addition to observational astronomy, astrophysics relies heavily on theoretical modeling and computer simulations to test hypotheses and make predictions about the behavior of celestial objects and phenomena. By combining observations with theoretical models, astrophysicists can gain a deeper understanding of the physical processes at work in the universe and make significant contributions to our understanding of the cosmos…

22 | THE COSMIC CHRONOLOGY

Chapter 4

What is Cosmology?

Cosmology is the branch of astronomy that deals with the study of the origin, evolution, structure, and eventual fate of the universe as a whole. It seeks to answer some of the most profound questions about the nature of existence, such as how the universe began. What is it made up of? In addition, how it will evolve in the future? Cosmologists investigate the large-scale properties of the universe, including its overall shape, size, and composition, as well as the processes that govern its behavior over time.

One of the central concepts in cosmology is the Big Bang theory, which posits that the universe began as a hot, dense, and infinitely small point roughly 13.8 billion years ago. According to this theory, the universe has been expanding and cooling ever since, giving rise to the vast array of galaxies, stars, and other cosmic structures that we observe today. Cosmologists study the remnants of the Big Bang, such as the cosmic microwave background radiation, to gain insights into the early history of the universe and test theories of cosmic inflation and the formation of cosmic structure.

Cosmology also encompasses the study of dark matter and dark energy, two mysterious substances that make up the majority of the universe's mass-energy content. While dark matter exerts gravitational influence on visible matter and helps to explain the structure of galaxies and galaxy clusters, dark energy is to be responsible for the observed accelerated expansion of the universe. Understanding the nature of dark matter and dark energy is one of the key challenges in cosmology, with profound implications for our understanding of the fundamental forces that shape the universe on the largest scales.

25 | THE COSMIC CHRONOLOGY

Chapter 5

Modern Instruments we use

We know that the modern era in space exploration has a lot to understand. Many mysteries face up to a challenging concept. Whether it might be a space mission or space exploration. We as humans use specific instruments to observe the night sky and the celestial objects. We even use other instruments to detect things. Let us know about the types of telescopes and the instruments used in cosmology and astrophysics. Telescopes and instruments play a crucial role in the field of cosmology, allowing scientists to observe and analyze celestial objects and phenomena across vast distances in space. Here is an overview of some key telescopes and instruments used in cosmological research:

Optical Telescopes: These telescopes use lenses or mirrors to gather and focus visible light from celestial objects. They are essential for observing stars, galaxies, and other luminous objects in the universe. Examples include the Hubble Space Telescope (HST), which has provided stunning images and data on distant galaxies, nebulae, and other cosmic phenomena.

Radio Telescopes: Unlike optical telescopes, radio telescopes detect radio waves emitted by celestial objects. They are crucial for studying phenomena such as pulsars, quasars, and the cosmic microwave background radiation (CMB), which provides important clues about the early universe. The Very Large Array (VLA) in New Mexico and the Atacama Large Millimeter/sub-millimeter Array (ALMA) in Chile are prominent examples of radio telescopes used in cosmology.

X-ray Telescopes: X-ray telescopes observe high-energy X-rays emitted by objects such as black holes, neutron stars, and hot gas in galaxy clusters. These telescopes provide valuable insights into extreme environments and energetic processes in the universe. Examples include the Chandra X-ray Observatory and the XMM-Newton satellite.

Infrared Telescopes: Infrared telescopes detect infrared radiation emitted by celestial objects, revealing details that are often obscured in visible light. They are essential for studying cool objects such as protostars, dust clouds, and distant galaxies. The Spitzer Space Telescope and the James Webb Space Telescope (JWST) are notable examples of infrared telescopes used in cosmology.

Now, after the overview of the telescopes, let us know about few instruments that we humans use in astrophysics and cosmology:

Particle Detectors: Particle detectors are crucial for studying high-energy particles generated by cosmic phenomena such as supernovae, black holes, and cosmic rays. Devices like particle accelerators, such as the Large Hadron Collider (LHC), allow scientists to recreate conditions similar to those in the early universe and probe the fundamental building blocks of matter.

Spectrographs: Spectrographs are instruments that break down light into its component wavelengths, revealing information about the chemical composition, temperature, and motion of celestial objects. They are used to analyze the spectra of stars, galaxies, and interstellar gas clouds, providing insights into their properties and evolutionary processes.

Imaging Cameras: Imaging cameras capture detailed images of astronomical objects. Across various wavelengths, from visible light to radio waves and beyond. These images serve as essential data for astronomers to study the morphology, structure, and dynamics of galaxies, nebulae, and other cosmic entities.

Gravitational Wave Detectors: Gravitational wave detectors, such as the Laser Interferometer Gravitational-Wave Observatory (LIGO) and the Virgo interferometer, detect ripples in spacetime caused by cataclysmic events such as black hole mergers and neutron star collisions. By observing gravitational waves, scientists can study phenomena that are invisible to traditional telescopes, providing new insights into the nature of gravity and the universe's most violent events.

Cosmic Microwave Background (CMB) Detectors: Instruments designed to observe the CMB, the residual radiation from the Big

Bang, provide crucial information about the early universe's properties and evolution. CMB detectors, such as those used in experiments like the Planck satellite and the Wilkinson Microwave Anisotropy Probe (WMAP), map the faint fluctuations in the CMB, offering insights into the universe's structure and composition.

In summary, instruments used in cosmology and astrophysics encompass a diverse range of technologies that enable scientists to explore the universe's mysteries across different scales and wavelengths. These instruments play a crucial role in advancing our understanding of the cosmos and uncovering its fundamental principles.

Listing all instruments used in cosmology and astrophysics would be an extensive task as the field encompasses a wide array of technologies across various wavelengths and scales. However, I can provide a comprehensive list covering some of the most prominent and commonly used instruments:

Optical Telescopes

Radio Telescopes

X-ray Telescopes

Infrared Telescopes

Gamma-ray Telescopes

Gravitational Wave Detectors (e.g., LIGO, Virgo)

Cosmic Microwave Background (CMB) Detectors (e.g., Planck, WMAP)

Particle Accelerators (e.g., Large Hadron Collider)

Spectrographs

Imaging Cameras

Neutrino Detectors

Cosmic Ray Detectors

Space-based Observatories (e.g., Hubble Space Telescope, Chandra X-ray Observatory)

Ground-based Observatories (e.g., Keck Observatory, Very Large Telescope)

Neutron Detectors

Dark Matter Detectors (e.g., DAMA, XENON)

Neutrino Observatories (e.g., Ice Cube, Super-Kamiokande)

Cosmic-ray Observatories (e.g., Pierre Auger Observatory)

Solar Observatories (e.g., Solar and Heliospheric Observatory, Solar Dynamics Observatory)

Interferometers (e.g., VLBI, ALMA)

This list represents a broad range of instruments and technologies used by scientists in their quest to explore and understand the universe. Keep in mind that the field is continuously evolving, and new instruments are constantly being developed to push the boundaries of our knowledge further.

33 | THE COSMIC CHRONOLOGY

Chapter 6
About the universe

The universe, an expansive and enigmatic realm, has captivated human curiosity for millennia. It encompasses all of existence, from the smallest subatomic particles to the largest cosmic structures, stretching across unimaginable distances and harboring untold mysteries. At its core, the universe serves as the stage for the unfolding drama of cosmic evolution, where galaxies collide; stars are born and die and the laws of physics govern the fabric of spacetime.

One of the most striking features of the universe is its vastness, extending far beyond the reaches of human comprehension. It comprises billions of galaxies, each containing billions of stars, along with countless planets, nebulae, and other celestial objects scattered throughout the cosmic expanse. The sheer scale of the universe is both humbling and awe-inspiring, challenging our understanding of space and time and prompting profound questions about our place within it.

Within this grand tapestry of cosmic wonders, the universe exhibits remarkable diversity and complexity. Galaxies come in various shapes and sizes, from majestic spirals with swirling arms to irregular conglomerations of stars and gas. Stars themselves undergo a lifecycle of birth, fusion, and eventual demise, shaping the evolution of galaxies and the distribution of elements throughout the cosmos. Nebulae, vast clouds of gas and dust, serve as the birthplaces of stars and the crucibles for the formation of planets and solar systems.

Moreover, fundamental physical laws and principles that dictate its behavior and evolution govern the universe. From the force of gravity, which shapes the structure of galaxies and drives the motion of celestial bodies, to the electromagnetic forces that govern the behavior of light and matter, these fundamental forces underpin the workings of the cosmos. Through the lens of modern astrophysics and cosmology, scientists strive to unravel the mysteries of the universe, probing its origins, composition, and ultimate fate.

Chapter 7

The traces of the beginning -1: The Big Bang

The universe is about 13.8 billion years. From a single point came the largest and unknown celestial body. The event when the universe first started to expand from a single point in known as The Big Bang. An astrophysicist, Fred Hoyle named Big Bang. As per our current technology, research and models, we only know things from when $t=10^{-43}$s from the Big Bang. As it is a very small, tiny, miniature period, we cannot really trace all the satisfying possibilities. Once such principal is:

Heisenberg uncertainty principle:

We cannot know both the position and speed of a particle, such as a photon or electron, with perfect accuracy.

$$\Delta x \Delta p \geq h/4\pi$$

When working with complimentary variables like energy, time, the more certainly associated with the less certainly that can be associated with the other. This allows quantum fluctuations, which is a very real and measureable phenomenon by which the particles pop into and out of existence all over the universe.

- Big Bang :
    Following forces came to existence:
    Weak nuclear force
    Strong nuclear force
    Gravitational force

The period from (t=0s) to (t=10^{-43}s), is called the Planck Epoch. The temperatures were still hot but slowly cooling. At this point, the temperature was 10^{32}k. We all know that the Quantum Gravity is also called the theory of everything. Since general relativity describes the large scale, or cosmological, structure of the universe and quantum theory describes the microscopic, or subatomic, structures, the unification of these theories would explain both the very big and the very small. This theory is often referred to as a "theory of everything". Now the quantum gravity has come to a sense.

From ($t = 10^{-43}$s) to ($t = 10^{-36}$s) is called the Grand Unification Epoch, temperature cooled down to 10^{29}k. Then, we had our four fundamental forces coupled with each other. Due to cooling down of temperature, gravitational force decoupled from the other. The other three were known as the Electro-strong force.

The period from ($t = 10^{-36}$s) to ($t = 10^{-32}$s), is known as the Electro Weak Epoch. This is when the Strong nuclear force decouples from the other two forces. The other two were known as the Electro Weak force. At this point, universe cooled down to 10^{28}k. Suddenly at this point universe expanded by 26 orders of magnitude. This is like going from the size of a molecule (less than billionth of a meter) to an object that is 10 light years across (60 trillion miles). This Expansion produced hot plasma of Quarks, Anti-Quarks and Gluons.

After (t=10^{-20}s), eventually (t=10^{-12}s), we enter the Quark Epoch. Temperature cooled down to 10^{12} K where electromagnetic force de-coupled with weak nuclear force. The Higgs field bestows particles with mass for the first time but things are still too hot for protons and neutrons to form. We can generate these energies in particle accelerator.

Theoretical → Experimental

At around (t=10^{-6} s) it was 10^{10} K. It is cool enough for Hadrons to form.

From (t=10^{-6}s) to (t=1s), it is known as Hadron Epoch. This has happened in just one second.

From (t=1s) to (t=10s), it is known as Leptons Epoch, where Hadrons and Anti-Hadrons largely annihilate, leaving Leptons and Anti-Leptons to dominate and annihilate to matter and anti-matter.

From (t=10 s) to (t=17 min), it is known as Big Bang Nucleosynthesis. Temperature was 10^9 K to 10^7 K. Cool enough for Baryons to be stable but also hot enough for them to fuse.

It was cool enough for fusion of. The mass of the hydrogen and helium was in the ratio of 3:1 (by mass). At this point, the universe is 600 light years across and expansion continued.

Over the next 337,000 years, all the helium nuclei collectively form clouds of gas. Now the General Theory of Relativity* came to sense (matter wraps spacetime). The universe slowly becomes structured.

*The General Theory of Relativity, developed by Albert Einstein in 1915, revolutionized our understanding of gravity and the universe. Here are the key points and formulas:

1. **Principle of Equivalence**: Gravity is not a force but the result of the curvature of spacetime caused by mass and energy.
2. **Curvature of Spacetime**: Massive objects like stars and planets curve the fabric of spacetime, affecting the motion of other objects.
3. **Geodesics**: Objects follow the shortest path (geodesic) through curved spacetime, which appears as curved motion in flat space.
4. **Einstein Field Equations**: The fundamental equations of general relativity describe the relationship between the curvature of

spacetime and the distribution of matter and energy. They are expressed as:

$$G_{\mu\nu} = 8\pi T_{\mu\nu}$$

Where:

- *$G_{\mu\nu}$ represents the Einstein tensor, describing the curvature of spacetime.*
- *$T_{\mu\nu}$ represents the stress-energy tensor, describing the distribution of matter and energy.*

5. **The Metric Tensor (g)**: Describes the geometry of spacetime. It defines the interval between two events in spacetime and is represented as $ds^2 = g_{\mu\nu} dx^\mu dx^\nu$.
6. **Schwarzschild Solution**: Describes the spacetime around a spherically symmetric, non-rotating mass. The metric is given by:

$$ds^2 = -\left(1 - \frac{2GM}{c^2 r}\right) c^2 dt^2 + \left(1 - \frac{2GM}{c^2 r}\right)^{-1} dr^2 + r^2 (d\theta^2 + \sin^2\theta \, d\phi^2)$$

Where:

- *M is the mass of the object.*
- *G is the gravitational constant.*
- *c is the speed of light.*
- *r is the radial coordinate.*
- *θ and ϕ are the angular coordinates.*

7. ***Gravitational Time Dilation***: *Clocks in strong gravitational fields run slower than those in weaker fields. The time dilation factor is given by*

$$1 - 2GMc2r1 - c2r2GM$$

These are the essential concepts and formulas of the General Theory of Relativity, providing a profound understanding of gravity and the structure of the universe.

During the next era, recombination and photons decoupling took place after about 377 thousand years from the Big Bang. Things have come to 4000K; cool enough for electrons to combine with nuclei to form neutral atoms for the first time. Electrons are captured down on the state of photons in doing so. This marks that the universe is visible, in the sense that we consider something visible to our eyes. It is no longer opaque, but transparent with electromagnetic radiation freely moving over large distances. Currently, at this point, universe diameter is 100,000,000 light years across.

For the next 150 million years, nothing much happened. This period is known as Dark Ages. There was a huge amount of hydrogen and helium and photons but no stasis. Slowly things cooled down to 300K that allowed for liquid water, if any were to exist. Then all way to 60K, which is a temperature that is finally cold enough for human standards to somewhat resemble the cold outer space. But, slowly ever so slowly all the hydrogen and helium gas continued to collect and over years, million years, the minute gravity exerted by these particles combined with the more significant gravity exerted by the surrounding Dark Matter , pulled matter towards into clumps. The dense it becomes, the more gravity is exerted. Gravity pulled atoms towards it until the atoms resist the other atoms to collide with them. This formed a bond between the matted but due to more atoms, there was an ignition.

45 | THE COSMIC CHRONOLOGY

Chapter 8

The traces of the beginning -2: Forming stars and galaxies in the young universe

As the energy molecules are converted into matter, Hydrogen becomes as what we see today on Earth.

From 150 million years to a billion years, all the Hydrogen and Helium clouds slowly start to accumulate into higher. As we learned in modern physics from Einstein's general theory of relativity, objects with mass warp spacetime, inducing a curvature that attracts all massive objects to each other. Even though the matter in the universe consisted only of tiny atoms at this time, these still exert gravity. So when the minute gravity slowly starts to attract things and make an accumulated region of a little higher density the usual, the gravity becomes little stronger and more matter accumulates and makes up a dense region as hydrogen did so. When these gasses accumulate, we know from the chemistry that particles within a sample of gas exert certain pressure or outward force. However, we said that matter exerts gravity hence; the outward force counters that gravitational force. This type of outward force countering gravity makes a state where the whole gas cloud is in Hydrostatic equilibrium. These gas clouds are very large that they stretch up to < 1 light year (Remember: 1 light year = 9.7 trillion km). A gas cloud across a light year or more is known as a Nebula, will remain in equilibrium if the kinetic energy of gas pressure, or the force pushing out, is precisely in balance with the gravitational force as in a hydrostatic equilibrium.

E_1 = kinetic energy of the gas pressure

E_2 = gravitational potential energy

Here, E_1 is countering E_2 = 0 Net force

There are few consequences, if the cloud exceeds the threshold called the Jeans Mass, which is a thousand times more massive than our sun, gravity dominates the kinetic energy, which results in gravitational collapse

$$M_{cloud} > \text{Jeans mass}$$

Any net rotation is amplified as the cloud is flattened into a disk by the centrifugal force, with gravity pulling matter towards the centre of the disk. The precise physics involved are complicated but makes the sense of subject short. Things started getting hotter and hotter as the matter keeps collapsing at the centre for over a millions of years, until the atoms are reionized back into plasma. With temperatures too hot for neutral atoms to exist, eventually the inner region of the gas was so hot that the outward force further collapsed and counters the gravity and at this stage we call it a Protonstar as it is in a temporary Hydrostatic equilibrium. More material from the surrounding collects on the Protonstar, increasing the gravitational potential energy, with sufficient additional mass, makes the gravitational force superior, the outward force becomes minimal, and that makes the matter to collapse evermore faster making the temperatures rise to a millions of degrees. Now started the nuclear fusion, during nuclear fusion the tremendous energy produced by the fusion generates more temperature with makes the body even more hot making the outer layers of the matter into plasma. Nuclear fusion counter the gravitational force as it produces outward force as it undergoes the process of nuclear fusion. That is how a star stops itself from further collapsing. That is how a star is born. While this depiction is overly simplified, it is very enough to imagine the events and understand. So with that let us define a star. A star is a gigantic gas cloud that collapsed into itself that with a tremendous gravitational force and it becomes a hot, spherical furnace of plasma. The heat then triggers nuclear fusion resulting outward pressure contracting gravity. This makes the star to be together and does not expand or contract.

The formation of a star takes millions and billions of years and this result in tremendous radiation, which triggers re-ionization in the surrounding networks of nebula, stripping these gas particles of their electrons. The precise nature of a star is dependent on its mass. Massive stars exert more gravity and when they collect at a dense region, they form the first galaxy. However, before going to galaxies we must know how to classify stars.

Chapter 9

The Classification of stars using spectrum.

Now let us look at some stars. We can observe a panorama of stars whirling around in galaxies, which have subsequently gathered into clusters and superclusters, at this point in the universe's history, roughly a billion years in the past. So what took place next? Let us study the fundamentals of this system now, as the solution to this question will need our learning more about stars and the traits that define how we classify them. We originally classified stars into colour classes when we began seeing them via telescopes.

Deep crimson, yellow, red, and white. Later on, this was improved, and every colour was divided into individual letters: A to D stood for white, E to L for yellow, and M and N for red. Although it was eventually discovered that classifying stars according to their surface temperatures made more sense, the letter system was kept in place since the classification of stars had previously been thoroughly researched. According to the

Harvard method of star classification, we currently have O, B, A, F, G, K, and M stars, which are the hottest at around 25,000 Kelvin and the coldest at about 3,500 Kelvin. Annie Jump Cannon (1863 – 1941), an early astronomer, developed this. Although this letter arrangement is a little odd, we may utilise the following mnemonic to help us remember it: Oh, kiss me and be a good girl! Feel free to swap out the female with the boy, whatever appeals to you more.

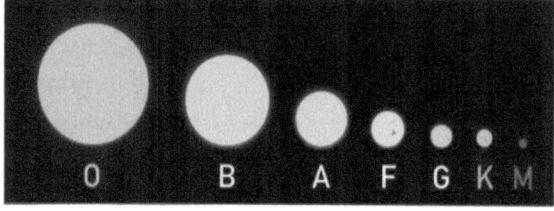

Source: https://www.google.com/url?sa=i&url=https%3A%2F%2Fwww.quora.com%2FIve-read-and-heard-the-term-M-class-star-before-but-what-are-the-other-classes-of-stars&psig=AOvVaw1pZ9ReBezEFnCxdfkti3xC&ust=1714124100741000&source=images&cd=vfe&opi=89978449&ved=0CBIQjRxqFwoTCIi9wI-I3YUDFQAAAAAdAAAAABAK

Similar to how we studied the Bohr model* in general chemistry, we analyse the light we get from stars and associate it with specific elements and temperatures. The more hydrogen and helium nuclei that have lost their electrons to create the phase of matter known as plasma, the hotter the star is.

O stars exhibit extremely little hydrogen because the majority of hydrogen is electron-deficient and unable to absorb or release light. We do observe helium-correlated emission as helium can still hold one or both electrons. The spectrum changes as hydrogen cools down somewhat with stars and can now hold onto one electron. As things cool down even more, bands corresponding to elements such as calcium appear. Not only does temperature determine convention, but it also has an effect on colour and size. K and M stars are red, whereas O and B stars, which are hotter, are bluer.

Because they are burning through so much more fuel, hotter stars typically have bigger, brighter burning regions. All of this temperature and luminosity data, together with some oblique mass and radius information, may be shown on a Hertzsprung-Russell diagram, or H-R diagram for short. This diagram's vertical axis represents brightness, or the amount of energy released by a certain star per unit of time, rising moving up, while

54 | THE COSMIC CHRONOLOGY

the horizontal axis shows temperature dropping to the right. It is evident that most stars lie on a continuous curve, which we refer to as main sequence stars. This pattern is followed by 90% of all stars, including our sun, which is a member of this golden area. Most stars are in this main sequence, although some are very cold and bright, like red giants, and some are very hot and faint, like white dwarfs. Though temperature and brightness are the only variables shown in this picture, we may still deduce a great deal about other factors.

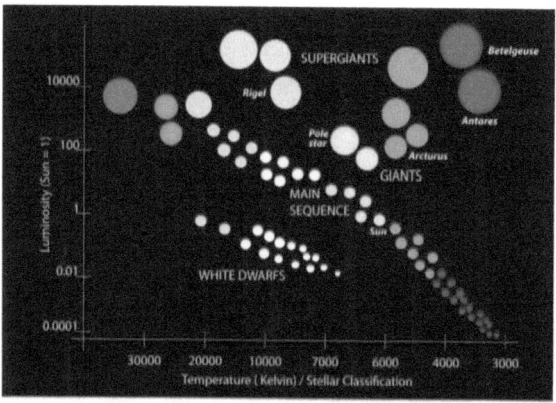

The Hertzsprung–Russell diagram

Source:
https://www.google.com/url?sa=i&url=https%3A%2F%2Fwww.azoquantum.com%2FArticle.aspx%3FArticleID%3D436&psig=AOvVaw2Cvj06Xzc1y9vjoC69tr81&ust=1714124036652000&source=images&cd=vfe&opi=89978449&ved=0CBIQjRxqFwoTCMiltOWH3YUDFQAAAAAdAAAAABAE

Because larger stars radiate more energy due to their larger surface areas, they are always more brilliant. Additionally, when we travel from left to right, we can plainly observe that colour correlates with temperature. The representation of size is also shown, with red giants and white dwarfs diverging from the general pattern of main sequence stars getting smaller from left to right. This information, which was gathered in the early 20th century by examining hundreds of thousands of stars, provides details on stars, including the previously mentioned mass-luminosity link.

*The Bohr model, also known as the Rutherford–Bohr model of the atom, was introduced by Niels Bohr and Ernest Rutherford in 1913 and describes how an atom is made up of an orbiting electron ringed by a tiny, dense nucleus. Its structure is similar to that of the Solar System. With the electron energies quantized (presuming only discrete values) and electrostatic force acting as the attraction instead of gravity. It came after and eventually supplanted a number of previous models in the annals of atomic physics, such as Joseph Larmor's Solar System model (1897), Jean Perrin's model (1901). The Saturnian model (1904) by Hantaro Nagaoka, the plum pudding model (1904). The quantum model (1910) by Arthur Haas, the Rutherford model (1911), and the nuclear quantum *model (1912) by John William Nicholson. The new quantum mechanical interpretation presented by Haas and Nicholson, which abandoned any effort to explain radiation in terms of classical physics, was the primary advance over the 1911 Rutherford model.*

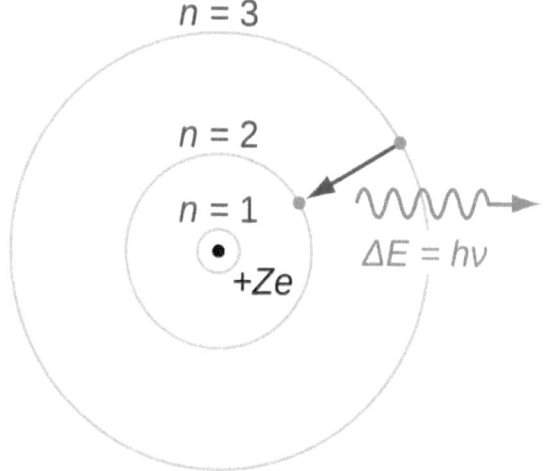

source:

https://www.google.com/url?sa=i&url=https%3A%2F%2Fen.wikipedia.org%2Fwiki%2FBohr_model&psig=AOvVaw3SsfOzeNoVLPTC3r44TEBy&ust=1714126652786000&source=images&cd=vfe&opi=89978449&ved=0CBIQjRxqFwoTCIDyotWR3YUDFQAAAAAdAAAAABAI

Therefore, to wrap up this chapter we must know the terminology to sort stars.

Overview:

According to Harvard method

O – Hottest

B, A, F – Hot

G, K, M – cold

According to the Hertzsprung - Russell diagram (H-R diagram)

Main Sequence:-

Blue stars: Big, Hot, Bright (up to 200 solar masses*)

Yellow stars: In between/ moderate (up to 1 solar mass*)

Red stars: Small, cool, dim (down to 0.1 solar mass*)

Red Giants:-

Red and cool (up to 0.3 – 8 solar mass*)

White Dwarfs:-

Tiny and hot (0.2 – 1.3 solar mass*)

*The **solar mass** ($M_\odot$) is a standard unit of mass in astronomy, equal to approximately 2×10^{30} kg. It is approximately equal to the mass of the Sun. It is often used to indicate the masses of other stars, as well as stellar clusters, nebulae, galaxies and black holes. This equates to about two nonillion (short scale), two quintillion (long scale) kilograms, 2000 quettagrams, or 2 quettakilograms:*

$M_\odot = (1.98847 \pm 0.00007) \times 10^{30}$ kg

The solar mass is about 333000 times the mass of Earth (M_E), or 1047 times the mass of Jupiter (M_J). The value of the gravitational constant was first derived from measurements that were made by Henry Cavendish in 1798 with a torsion balance. The value he obtained differs by only 1% from the modern value, but was not as precise.

THE COSMIC CHRONOLOGY

Chapter 18
Over to the end of the stars

We have gone through how the stars are born and how we classify them into different forms. Diversity indeed shows the miracles of the universe. So let us know how stars die. In typical situations, we must first know the periodic table. You can refer this table given below at anytime.

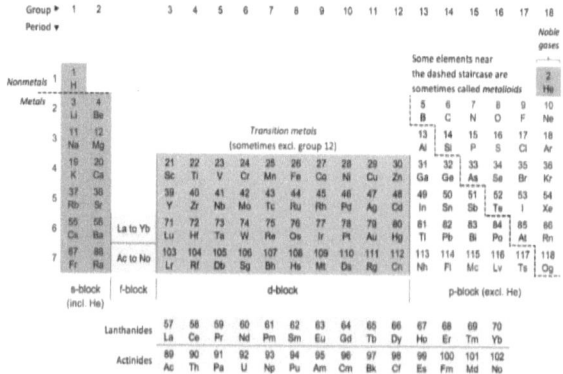

Source: https://upload.wikimedia.org/wikipedia/commons/8/89/Colour_18-col_PT_with_labels.png

Now a day we see many elements in our daily life. It might be in our body or in the earth. However, we only know that Hydrogen and Helium were the first elements to come into this universe and fuel up the stars. What about the other? Would you believe that if I said, all the elements are formed when the stars die?

We now possess the vocabulary required to characterise and classify the multitude of stars and galaxies that result from our understanding of what transpired during the first billion years or so of the universe's existence. However, since we have only discussed hydrogen and helium thus far, we have not yet discussed all the other elements on the periodic table. Where did the other elements originate? Moreover, all the planets and moons together? How did they arrive here? Yes, it is true that stars do die. The manner in which they die and what remains thereafter will vary depending on their mass.

To prepare us for the next 13 billion years of cosmic evolution, let us take a look at the lives of a few different types of stars. Any star's life cycle, from conception to expiration and all phases in between, will take millions or maybe billions of years. Because of this, stars appear to remain constant, even though a human lifetime is but a tiny fraction of an eyeblink to these enormous creatures. Since gas serves as the star's fuel, the mass of a given star—the amount of gas that gathered and condensed to form the star—determines the star's trajectory. As we might recall from chemistry and physics, According to

$$E = mc^2$$

Energy = Mass X speed of light (3.00 x 10^8 m/s)

Nuclei fuse when they collide with enough energy to overcome the electromagnetic repulsion between them. A little portion of their mass is converted into enormous amounts of pure energy during this process.

A star can only produce enough outward energy to offset the effects of gravity's relentless inward squeezing by hitting and fusing nuclei in its extremely hot core. This implies that the quantity of matter that makes up the star controls the amount of fuel, as well as the star's lifetime and ultimate destiny through a number of other parameters. Firstly, let us discuss a low-mass star. This would vary from a star. That is roughly the mass of our sun to the tiniest stars that can exist. This refers to the least quantity of material that may adequately ignite nuclear fusion to be classified as a star, which are around thirteen Jupiter masses. Any star will start as a cloud of gas and dust that is at least a few light years wide, as we already know. Since this was all that was left over from the brief seventeen minutes of Nucleosynthesis that occurred shortly after the Big Bang, hydrogen and helium made up the majority of this material in the early stages of star formation. Gravity causes this matter to gather and push inward as it contracts. After a few million years, the temperature rises to the point where nuclear fusion finally starts, creating equilibrium and a main sequence star that glows yellow or red from the energy released from the internal collisions. These fusion processes start with the fusion of two protons, which is followed by beta decay to produce a proton and a neutron. This is known as a deuteron, and it is the heavy hydrogen nucleus. Helium, which contains two protons and two neutrons, is then produced by processes involving deuterons. This is how a star like this will persist for billions of years, gradually converting all of the hydrogen in its core to helium while keeping its size, temperature, and brightness mostly constant until nearly all of the hydrogen is gone. Things start to truly alter at this time. The star's core will contract and heat up, causing the remaining hydrogen to burn even more quickly. Because of all the additional energy being produced, the outer layers of the star will be pushed away from the core by radiation. The star starts to pulse as it exhausts its last source of energy and enters what is known as the horizontal branch. During this phase, it gets smaller, hotter, and bluer until finally a significant portion of the helium has been combined into bigger nuclei. The star will have very little material left to burn after the core is mostly composed of carbon and oxygen, with just a shell of helium and a shell of hydrogen around it. As a result, the core will collapse and the star will reach the asymptotic giant branch. This indicates that it will continue to expand quickly and reshape into a massive star until the last energy outbursts drive the outer layer away from the core and back into the interstellar medium, leaving behind only a small, extremely hot, naked core that is roughly the size of Earth. It cannot burn any more fuel and is not hot enough to fuse oxygen or carbon atoms, therefore it will progressively cool and compress until it is reduced to the size of a white dwarf star. Since the expelled shell is not a planet and did not originate from one. The term "planetary nebula" is deceptive; yet, uncertainty surrounding its discovery led to its adoption. After then, the material within a planetary nebula will open up to combine with other gas particles to produce an additional star. Things are very different with high-mass stars, which are stars with masses significantly greater than our sun. Their death won't be so peaceful. Stars leave on a high note. Ordinarily, things begin with a cloud of gas accumulating due to gravity. It just means that this cloud will have a lot more mass because it will be larger than those that form low-mass stars. Increased mass results in increased gravity, which increases the force pressing inward and raises the star's temperature. Faster fusion at a hotter temperature results in increased outward pressure, which balances the increased inward attraction of gravity. This will produce a hot, massive, brilliant, blue, main-sequence star.

This is the point at which the situation diverges from that of low-mass stars. High-mass stars burn their fuel far more quickly because they are much hotter than low mass stars, which take billions of years to exhaust their fuel. This implies that they exhaust all of the hydrogen in their cores in a matter of only a few hundred million years, or ten million if they are large enough. Similar to low-mass stars, the star will enlarge into a massive star as its fuel runs out because the core of the star constricts and warms up, generating more energy. As a high-mass star's core compresses, it becomes significantly hotter than a low-mass star's core, allowing it to fuse helium nuclei to form carbon, oxygen, neon, and finally silicon. Each heavier nucleus is then limited to a smaller and smaller area of the core that is hot enough to fuse. Iron, the heaviest element that can combine within a star, is located all the way in the centre. The star is left with a core of iron nuclei that are so stable that more fusion cannot release any more energy as this happens in these several layers, each of which performs a specific sort of fusion until no fuel is left. The star collapses in a split second when gravity takes over at this moment, ejecting all of the heavy nuclei it has produced into space as the outer layers bounce off the core and cause an explosion. A supernova is this amazing explosion, one of the universe's most violent and energetic phenomena. A supernova produces an incredible surge of energy that can also be used to synthesise hundreds of elements heavier than iron in this brief instant. Any element with an atomic number higher than either 26 is created in a supernova, a rare occurrence such as the collision of two neutron stars, or between a neutron star and a black hole, which are objects we cannot see. This includes nickel, copper, zinc, silver, and gold. Because stars are unable to synthesise these heavy elements in the same way that they can synthesise all the elements up to iron throughout the course of their long lifespan, they are far more rare than elements like carbon and oxygen. Only in select unusual collision situations or during the demise of a high-mass star can nature produce these uncommon elements. Supernovae can also be seen with the naked eye in our own galaxy, such as the one that produced the well-known Crab Nebula, which was documented by several civilizations in 1054. Supernovae are so bright that, when viewed through telescopes, they outshine the entire galaxy to which they belong. A white dwarf is not left behind by a supernova nowadays. White dwarfs are left behind by lower-mass stars that have a mass of less than eight solar masses in the beginning because there is insufficient gravity to resist electron degeneracy pressure once the star has shrunk to its lighter, earth-sized core. Put differently, a white dwarf will essentially merge into one massive metallic solid as the electron clouds surrounding its nuclei press against one another to stop it from collapsing any more. Even so, one teaspoon weighs about fifteen tonnes, demonstrating how dense this item is. The core of a star thus meets this fate below about 1.4 solar masses, the maximum mass of a white dwarf, also referred as the Chandrasekhar limit. The core of a star suffers from this. One of two things will however remain when a high-mass star dies, if its core is large enough for a supernova to occur and it is beyond the Chandrasekhar limit.

The supernova is caused by the shockwave from the collapse of the core. Which cannot support itself against gravity and collapses with such force that all the electrons are squeezed into protons and combine to form neutrons if the core is between about 1.4 and 3 solar masses, having been generated by a star that was originally somewhere in the ballpark of ten to forty solar masses. The result is a ball of entangled neutrons, like a giant atomic nucleus, that contains all of the mass that was originally in the star's core. A teaspoon of the star would weigh ten million tons. But what's even more amazing is that if the star's core is more than three solar masses, the outward pressure of the neutrons pressing directly against each other or the neutron degeneracy pressure isn't enough to stop the massive gravity. The neutrons will be smashed together as the rest of the mass collapses into a point of infinite density: the entire mass of a star's core, contained within a single point of zero volume. This is what we call a black hole. The outer layers of a star that has been blown apart, filled with heavy nuclei fused over the life of the star and the additional heavy nuclei formed during the star's supernova, will be called a nebula. However, the singularity that remains is anything but a nebula. Because of its infinite density, a black hole warps spacetime in such a way that light cannot escape. No matter what a black hole looks like, if it means anything at all, we'll probably never know because it's impossible for photons to escape it and get to our eyes, because that's how we look at things. It's unbelievable, but that's how nature works. Black holes do exist, and they're everywhere in the universe, like the remains of giant dead stars. They're so cool that they deserve their own chapter, which we'll get to later, but for now, let's take a look at what we've just learned about the life of a star. When a star forms from a massive cloud of gas, which is almost always between one-tenth the mass of the Sun and about thirty the mass of the Sun, the star is born somewhere on the main sequence. As the core runs out of fuel, it contracts, increasing the pressure around the core and pushing the outer layers outward, where they then cool to form a red giant. So all stars have a red giant phase when they are almost out of fuel, regardless of their mass. Eventually, when the star can no longer perform enough nuclear fusion to resist the effects of gravity, the star collapses, leaving behind a white dwarf if it has a low mass, a neutron star if it has an intermediate mass, and a black hole if it has one. it has a low mass. It has a particularly large mass. As we mentioned, black holes are some of the most fascinating objects in the universe and are a popular field of study for astronomers and theoretical physicists because there is so much we still do not understand about these strange creatures. Let us go ahead and learn a little more about black holes.

THE COSMIC CHRONOLOGY

Chapter 11

Discovering the Undiscovered: Black holes

We've just learned about what happens to very high-mass stars when they die. Once one of these runs out of fuel in its core, left with lots of iron and little else to fuse, there is nothing preventing gravitational collapse any longer. The outer layers plummet inwards in a single second, overcoming electron degeneracy pressure, and even neutron degeneracy pressure, generating a shock wave that triggers a supernova, and leaves behind a black hole. A single point containing most of the mass of the star. As much as this sounds like science fiction, we have mountains of direct evidence for these objects, even within our galaxy, so let's learn some more about these things and what they do. First of all, why are black holes black? To answer this, let's highlight some basics regarding escape velocity. As we may remember, general relativity tells us that massive objects warp spacetime, and the greater the mass, the more pronounced the influence. This is why objects fall down from the sky to the ground, they are simply following the curvature of space that is produced by the mass of the very nearby earth.

But something can escape earth's gravitational pull if it can travel away from the earth at a velocity that exceeds the earth's escape velocity.

$$v_e = \sqrt{\frac{2GM}{R}}$$

This means it is traveling so fast that it can climb out of the curvature, kind of like a ball rolling fast enough to get all the way up a hill. We can calculate the escape velocity for a massive object by using this equation, where G is the gravitational constant, M is the mass of the object, and R is the radius of the object. This is because the increasing density of the object will result in a more profound curvature of space. In the limit of a massive object with zero radius, the escape velocity becomes infinite, meaning that no object can escape the gravity of a black hole, no matter how fast it travels, even light, which travels at the speed of light, the fastest anything can go, as we learned when we talked about special relativity in the modern physics course. If light can't leave an object to reach our eyes, then we can't see it, and that's why black holes are black. This brings us to an interesting conclusion. Anything that is sufficiently dense so as to have an escape velocity greater than the speed of light must therefore be a black hole. Taking our previous equation and plugging in c for V, and then solving for R, we can plug in any mass we want into this equation and solve for the radius within which that object must be compressed in order to generate a black hole. This is called the Schwarzschild radius of an object. When a high-mass star collapses at the end of its life, it is compressed well beyond its Schwarzschild radius, which is the most common way that the universe produces a black hole. But hypothetically, anything could become a black hole if sufficiently compressed. Of course, the amount of compression required is staggering. For the earth to become a black hole, it would need to have a radius of just under a centimeter. That's all the mass of the entire world contained within something less than the size of a gumball. Even a person like you or me could become a black hole, but we would have to be compressed into a sphere with a radius of ten to the negative 25 meters, which is as much smaller than an atom as an atom is smaller than a person. So it's not such an easy task, which is why we don't see black holes in our day-to-day lives. So we know that all the material in a black hole is contained within its Schwarzschild radius, and there is a related term for describing this radius, which is called the event horizon. This is the region of spacetime surrounding the black hole within which light can't escape, and for a non-rotating black hole, the distance to the event horizon is equal to the Schwarzchild radius. We can consider this the boundary of the black hole in a certain sense, not because it's the edge of the matter producing the black hole, but rather because beyond this edge, spacetime is sufficiently warped such that light can't leave it and get to our eyes. But if we can't see black holes, how do we know they exist? Well there are many ways to observe black holes indirectly. Sometimes one star in a binary system becomes a black hole, and material from the other star begins accreting around what appears to be absolutely nothing. This activity emits X-rays that we can receive, which helps us identify that a black hole is there. More recently, we have been able to detect gravitational waves emitted from a pair of black holes that are merging.

Sometimes we can even see large collections of stars movingvery fast around a region of seemingly empty space. By measuring their velocities we can calculate the mass of what they must be orbiting, and it will typically be many solar masses, and that's a black hole too. Sometimes it will have a mass equal to many millions of solar masses, and we call this a supermassive black hole. These can form over billions of years if black holes swallow up enough material from their surroundings, or even merge with other black holes, and we have very convincing evidence that suggests there is a supermassive black hole at the center of every large galaxy in the universe, including ours. So what can possibly stop black holes from swallowing up everything in the universe? Well, for one thing, space is huge. Objects very close to a black hole may be doomed, but if you get far enough away from one, its gravity is no different than if it was a regular object. In other words, if our sun became a black hole today, we would certainly be in trouble in the sense that we wouldn't receive any more light and heat, but surprisingly, earth's orbit wouldn't change at all. The black hole would still have the same mass as the sun, so orbits would remain unperturbed. Another interesting feature of black holes was predicted by physicist Stephen Hawking. Against all intuition, it was found that black holes actually emit radiation, which we refer to as Hawking radiation. This can be understood by recalling the Heisenberg Uncertainty Principle, which allows for particle-antiparticle pairs to appear out of the quantum foam. One possible explanation claims that when such a quantum fluctuation occurs right at the event horizon of a black hole, one of these particles will fall into the black hole while the other escapes, and for energy conservation, the one that fell in will be the one with negative energy. This ever so slightly reduces the mass of the black hole.

This process is perpetual but painfully slow, with a solar-mass black hole requiring around 10 million to the 67 million years to completely evaporating. But, it does mean that even black holes eventually die, just like everything else in the universe. So with that, beyond having learned how to make black holes and super massive black holes, we now know enough about stars to start learning about the types of stars that have existed throughout the lifetime of the universe so far, and the structures that they have come together to form.

Chapter 12

Galactic Odyssey: Exploring the Cosmos

We've gained extensive knowledge about stars, encompassing their inception, evolution, and demise. Now, it's imperative to delve into the intricacies of their spatial distribution, spanning from microcosmic scales to grand galactic realms, while also uncovering additional insights into galactic dynamics and interactions. Let's rewind to a pivotal epoch, spanning from 150 million to one billion years post-Big Bang, when the primordial stars ignited. What distinguishes these celestial pioneers? Initially termed population three stars, this classification may seem paradoxical, given their status as the universe's inaugural stellar inhabitants. However, this designation stems from early categorization methods. Astronomers gauged stars' metallicity—essentially, the abundance of heavy elements. Those boasting substantial metal content were categorized as population one, while those with intermediate levels were labeled population two, and those with meager metal content were tagged as population three. We now grasp that low metallicity denotes an older star, as heavy elements only proliferated in the interstellar medium following the demise of the first-generation stars, dispersing their remnants. Thus, the inaugural stars, comprising solely hydrogen and helium from early universe nucleosynthesis, fall under population three, whereas population one stars, boasting comparatively higher metallicity, formed more recently, capitalizing on pre-existing heavy-element reservoirs within their gas and dust nebulae.

Moreover, comprehending that collapsing gas clouds frequently fragment during accretion, stars typically form within compact clusters. These stars often exist in gravitationally bound pairs, constituting binary systems, or in larger, multi-star configurations. Even more expansive congregations of stars are termed star clusters. Remarkably, most high-mass stars inhabit such systems, with solitary star systems being more prevalent among lower-mass stars like our Sun. Consequently, the iconic binary sunset portrayed on Tatooine in Star Wars might reflect a more commonplace occurrence in the galaxy than the solitary sunsets witnessed on Earth. It's worth noting, however, that extremely low-mass stars, such as red dwarfs, the most abundant type in our galaxy, typically exist in isolation. Regarding the stellar compositions within these clusters, they can comprise nearly any combination. One intriguing pairing involves a binary system comprising a white dwarf and a main-sequence star.

Zooming out from these smaller star systems, stars are typically ensnared within vast structures termed galaxies. These entities typically harbor anywhere from a few hundred million to several hundred billion stars, constituting the bulk of stellar populations, albeit with some adrift in the cosmic void between them. Galaxies manifest an array of sizes and shapes, as initially discerned by Edwin Hubble. He categorized some as spiral galaxies, characterized by thin, rotating stellar disks with prominent spiral arms extending outward. These structures host the majority of stars found in their outer regions. Others are labeled elliptical galaxies, presenting a smooth, featureless appearance distinct from the spiral morphology. Lastly, irregular galaxies defy classification into the aforementioned categories. These three primary galaxy classifications—abbreviated as S, E, and Irr—also feature subcategories. Spiral galaxies may include barred spirals denoted by SB, distinguished by spiral arms extending from a central bar rather than the galactic center. There are also galaxies designated as S0, featuring a thin disk devoid of spiral arms. Furthermore, true spirals can be further subdivided into Sa to Sd based on the tightness or looseness of their spiral arms, while ellipticals are categorized as E0, E2, E5, and E7 based on their spherical or flattened shapes.

Apart from exhibiting a diverse array of sizes, galaxies also differ in the stellar compositions they harbor. Elliptical galaxies predominantly host older population two stars, whereas spiral galaxies feature a blend of population two and population one stars, indicative of ongoing star formation fueled by high-density regions of gas and dust within their spiral arms. The formation of galaxies mirrors that of stars, with gravitational forces facilitating the aggregation of gas clouds into vast conglomerations, akin to the processes underlying star formation. Observational astronomy grants us the privilege of glimpsing early galaxy formations near the observable universe's edges, offering insight into the intricate cosmic ballet spanning billions of years. Some galaxies feature central quasars, initially dubbed quasi-stellar objects, characterized by supermassive black holes ensconced within galactic nuclei. These colossal entities, surrounded by accretion disks of gas, emit prodigious amounts of energy, rendering quasars thousands of times brighter than entire galaxies, facilitating their detection across vast cosmic distances.

Collisions and mergers play pivotal roles in shaping galactic evolution. Interactions between galaxies can induce distortions in shape and even culminate in total mergers, often referred to as galactic cannibalism. These events, while transformative, typically do not entail stellar collisions due to the vast expanses of space between individual stars. Galactic mergers, along with the gravitational influence of dark matter, significantly influence galactic morphology and evolution, giving rise to the diverse array of galactic structures observed today. Our understanding of the universe's history, from star and galaxy formation to the present day, hinges on fundamental principles governed by gravity—the driving force underlying all cosmic processes.

With a comprehensive overview of the universe's evolutionary trajectory thus far, the time has come to explore our cosmic abode. Amidst the myriad galaxies adorning the celestial canvas, which one is our home? Let's embark on this cosmic odyssey to uncover humanity's place in the vast expanse of the cosmos.

Chapter 13
Trails of the Milky Way

We've delved into the intricacies of star and galaxy formation, unraveling the mysteries of the vast universe. Among the countless galaxies, where does our home, the Milky Way Galaxy, reside? Exploring the Virgo Supercluster, containing smaller clusters like the Virgo Cluster and the Local Group, we focus on the latter. The Local Group harbors over fifty galaxies, ranging from dwarfs to giants. The Andromeda Galaxy stands as the largest, followed by our Milky Way Galaxy, a typical barred spiral galaxy housing 200 billion to 400 billion stars. Spanning 100,000 light-years in diameter and about a thousand light-years thick, the Milky Way features a disk with arms of varying sizes, encircled by a stellar halo, possibly housing a supermassive black hole at its core, similar to other galaxies.

A fraction of the Milky Way's visible mass comprises interstellar gas and dust, interspersed among stars. Noteworthy formations include open clusters like the Pleiades and distant globular clusters, hosting ancient population two stars. Additionally, the Milky Way hosts satellite galaxies like the

Small and Large Magellanic Clouds, engaging in gravitational interactions. Notably, the Milky Way and the Andromeda Galaxy are on a collision course, projected to merge in about four billion years, forming "Milkdromeda."

Examining the Milky Way's formation, rotational dynamics shape the spiral arms as stars orbit the galactic center. The oldest stars, population two, populate the halo, indicating early formation. Meanwhile, the disk comprises younger population one stars, witnessing ongoing star formation. The Milky Way's future entails stellar evolution and eventual demise, including events like the formation of our solar system around five billion years ago.

That was a very small and short chapter. Is this chapter is briefly described; it has a lot of depth if dived into.

THE COSMIC CHRONOLOGY

Chapter 14

The Formation of the Solar System

We've gained extensive knowledge about the formation of stars and galaxies, including our very own Milky Way galaxy. Now, let's delve deeper into our position within this vast cosmic structure. If we focus on a specific region within the Orion arm, located at a considerable distance from the galactic center, we encounter a yellow main sequence star. This star isn't particularly remarkable in terms of its size, boasting only 1 solar mass, making it relatively modest in stellar standards. However, its significance lies in the fact that it's our star, the one around which our planet orbits, commonly known as "the sun."

The sun, classified as a population one star, came into existence roughly 4.6 billion years ago from a cloud of gas and dust enriched with heavy elements expelled into interstellar space during the demise of older stars, particularly population three and two stars undergoing supernovas. This cloud, comprising hydrogen, silicates, iron, water, and various other compounds, underwent a process of spinning and flattening into a disk, akin to the formation process observed in galaxies, termed a protoplanetary disk.

Gravity played a pivotal role as the bulk of this cloud coalesced at the center, triggering fusion reactions and giving birth to the sun. Concurrently, the remaining material dispersed across varying distances from the rotational axis. Over time, minute dust particles began to collide and aggregate, forming larger rocks, which further collided to create planetesimals. These growing planetesimals exerted gravitational pull, attracting other nearby objects. Across hundreds of thousands of years, this cycle of collision and accumulation persisted, generating significant heat and maintaining molten conditions.

As debris cleared at specific orbital distances, these objects acquired sufficient mass to adopt a spherical shape under their gravitational influence, leading to the formation of the inner rocky planets observed today. Conversely, the gas and dust constituting the outer regions coalesced into spheres due to gravitational forces, containing substantial amounts of ice due to lower temperatures at greater distances from the sun. Upon reaching critical mass, these spheres attracted surrounding gases, evolving into the gas giants we recognize. Unsuccessful planetesimals either formed moons or persisted as smaller celestial bodies like asteroids and comets, with additional loose material gathering into rings encircling the larger planets, culminating in the birth of our solar system.

Since its formation, the solar system has undergone dynamic changes, with celestial objects frequently displaced. For instance, a few billion years ago, the alignment of Jupiter and Saturn caused alterations in Neptune's orbit, resulting in the migration of smaller objects from the outer solar system towards the inner planets, a phenomenon known as the Late Heavy Bombardment.

Moving forward, let's focus on the sun itself. Classified as a typical G star, or a yellow main-sequence star, the sun belongs to the population one category, denoting its relatively recent formation from a gas and dust cloud enriched with materials from the demise of older stars. The sun's outermost layer, the photosphere, burns at approximately 6000 Kelvin, while its dense inner core reaches temperatures of around 15 million Kelvin, maintaining a plasma state due to intense heat and gravity's crushing force.

Between the core and photosphere lies the radiative zone, where photons undergo absorption and emission in random directions before reaching the surface after a hundred thousand years. Below this lies the convection zone, where material cools as it ascends, subsequently sinking, heating up, and perpetuating a cyclical motion.

Surprisingly, the sun possesses an atmosphere consisting of the relatively cool chromosphere and the corona, which inexplicably exhibits temperatures soaring to a million Kelvin. While the exact mechanism behind this extreme heat remains elusive, it's widely believed to involve the sun's magnetic field, with some suggesting acoustic energy as a potential driver.

The sun's surface features dark patches known as sunspots, where magnetic field lines inhibit gas rise, resulting in cooler, darker regions. Additionally, prominences, plumes of plasma, and solar flares, eruptions of hot plasma in the chromosphere, are generated by the sun's magnetic field, contributing to solar activity.

Beyond these phenomena, solar activity generates the solar wind, a continuous stream of plasma and high-energy charged particles dispersing in all directions through space. This solar wind extends to the heliosphere's boundary, marking the sun's influence limit and, consequently, the solar system's edge, situated far beyond the outermost planet.

In comparison to the Milky Way, the solar system appears minuscule. If the galaxy were scaled to the size of the Earth, the solar system would resemble a pancake in proportion. Despite its diminutive size within the galactic context, the sun's sheer magnitude is undeniable. With a diameter over a hundred times greater than Earth's, it would require over a million Earths to fill its vast expanse. Furthermore, accounting for approximately 99.86 percent of the solar system's mass, the sun exerts significant gravitational influence, dictating the nearly circular orbits of most planets.

Transitioning to a brief overview of the planets, let's commence with Mercury, the sun's closest and barren neighbor. Next in line is Venus, characterized by its scorching temperatures and volcanic activity, rendering it a hostile environment. Our planet Earth follows suit, serving as the familiar abode for humanity. Subsequently, Mars, the red planet, awaits exploration, potentially heralding human endeavors beyond Earth's confines. Moving outward, we encounter the asteroid belt, marking the boundary between the inner rocky planets and the outer gas giants.

Continuing the planetary journey, Jupiter, the largest planet, reigns supreme, followed by Saturn, renowned for its captivating ring system. Further out lie Uranus and Neptune, the ice giants completing the roster of gas giants. Comprising four rocky and four gaseous planets, each accompanied by an array of moons, the solar system teems with celestial marvels awaiting further exploration and discovery.

As we reflect on the formation and composition of the solar system, it's evident that despite its complexity, its genesis can be attributed to fundamental astronomical processes. The fusion of heavy elements within high-mass stars and the subsequent dispersal of these elements into interstellar space during supernovas, coupled with the accretion of interstellar material into protoplanetary disks, lay the groundwork for solar system formation.

Moreover, this realization lends depth to our cosmic contemplations, underscoring humanity's intimate connection to the universe. Every atom comprising our being, excluding hydrogen, originated from the remnants of stars, journeying across space to contribute to the formation of the solar system. As Carl Sagan eloquently stated, we are "star stuff," imbued with cosmic origins that instill a profound sense of wonder and belonging as we gaze upon the celestial tapestry above. Lets dwell to know about the planets of the solar system.

THE COSMIC CHRONOLOGY

Chapter 15

The Messenger Planet: Mercury

Now that we possess a comprehensive comprehension of the origins of the universe, the Milky Way, and the solar system, along with a concise narrative of humanity's progression in understanding the solar system, we can embark on a detailed exploration of each constituent of the solar system. We categorize the eight planets into two sets of four: the inner quartet, known as terrestrial planets, characterized by their compact, rocky composition, and the outer quartet, known as gas giants, distinguished by their substantial gaseous structure. Beginning with the sun, let's systematically examine each celestial body outward.

Mercury, the closest planet to the Sun, is the smallest planet in our solar system. Mercury, the smallest planet in our solar system, boasts a radius merely one-third that of Earth's and a mass just one-twentieth as much. Its surface is a desolate expanse, characterized by a stark gray landscape pockmarked with craters. Proximity to the sun, with an average orbital radius of under 60 million kilometers, precludes any possibility of an atmosphere. Consequently, Mercury experiences extreme temperature differentials, soaring to a scorching 700 Kelvin during daylight hours and plummeting to a frigid 100 Kelvin at night, as there's no atmospheric insulation to retain heat. This extreme thermal dichotomy renders Mercury one of the hottest and coldest locales in our solar system.

Its name, Mercury, originates from its swift orbital velocity, observed by ancient astronomers who likened its movement across the sky to the rapid delivery of messages by the Roman god Mercury, the divine messenger. Despite our limited direct exploration—confined to flybys by Mariner 10 in the 1970s and the MESSENGER orbiter in the early 21st century—data obtained from these missions, coupled with density and gravitational calculations, have provided insights into Mercury's interior. Current understanding posits a dense iron-nickel core enveloped by a silicate crust, with the core constituting a substantial portion of the planet's volume, approximately 55 percent. Notably, this molten core generates a weak magnetic field, just 1 percent the strength of Earth's. However, Mercury's rotation is sluggish, completing a single axial rotation every 59 Earth days, resulting in a spin-orbit resonance where two Mercury years coincide with precisely three axial rotations.

As an "inferior planet," orbiting closer to the sun than Earth, Mercury's visibility in the sky is limited, presenting challenges for observation. However, its orbital peculiarities proved pivotal in advancing our understanding of gravity. The discrepancy in Mercury's orbit, a slow precession not accounted for by Newtonian mechanics, was reconciled by Einstein's general theory of relativity. This theory, elucidating the curvature of spacetime induced by massive objects like the sun, accurately predicted Mercury's orbit, resolving a longstanding astronomical puzzle.

Now, having delved into the enigmatic world of Mercury, our exploration of the terrestrial planets continues with a closer look at Venus.

THE COSMIC CHRONOLOGY

Chapter 16

The Evening Star: Venus

To delve deeper into the intricacies of Venus, let's first explore its proximity to Earth. Positioned as the second planet from the sun, Venus maintains an average orbital radius of approximately 108 million kilometers. Its relative closeness to Earth has earned it the moniker of Earth's sister planet, owing to similarities in size and mass. Once considered a potential twin to Earth, early speculations suggested Venus might harbor conditions conducive to life. However, reality starkly contrasts these hopeful conjectures, unveiling Venus's true nature as an inhospitable and formidable world.

Deriving its name from the Roman goddess of love, Venus presents an environment devoid of affection for prospective visitors. The planet boasts an exceedingly dense atmosphere, predominantly composed of carbon dioxide, exerting pressures nearly a hundred times greater than those experienced on Earth's surface. While terrestrial concerns regarding rising carbon dioxide levels persist, Venus showcases a stark contrast, with this gas constituting a staggering 96 percent of its atmosphere. This concentration, coupled with Venus's greenhouse effect, manifests in surface temperatures soaring to a scorching 735 Kelvin, eclipsing even Mercury's heat, despite the latter's closer proximity to the sun.

Beyond its atmospheric density, Venus cloaks itself in clouds laden with sulfuric acid, veiling its surface from prying telescopic gazes. Unlike the aqueous vapor clouds adorning Earth's skies, those enshrouding Venus emit a radiant brilliance, reflecting sunlight with remarkable intensity. This luminosity renders Venus a prominent fixture in both day and night skies, earning it epithets such as the "evening star" or "morning star" throughout antiquity.

Early attempts at uncovering Venus's surface characteristics yielded minimal success until the advent of space exploration. Pioneering missions such as the Pioneer Venus Orbiter and the Venera landers furnished humanity with vital insights into Venus's geology, culminating in detailed surface mapping courtesy of the Magellan orbiter. These endeavors revealed a predominantly flat terrain punctuated by elevated regions akin to terrestrial continents, christened Ishtar Terra and Aphrodite Terra in homage to love deities of varying cultures.

Further scrutiny exposed signs of recent volcanic activity, hinting at a dynamic geological history distinct from Earth's. The planet's iron-rich core, analogous to Earth's, belies its most striking divergence—the contrary rotation direction. While Earth and its terrestrial counterparts rotate in a prograde fashion, Venus defies convention, spinning retrograde with the sun rising in the west and setting in the east. This anomaly, attributed to a primordial collision altering Venus's rotational dynamics, remains a subject of ongoing scientific inquiry.

With Venus's intricacies laid bare, we embark on a journey closer to home, transitioning our focus to Earth—our cherished abode—and unraveling its origin, composition, and unique characteristics.

103 | THE COSMIC CHRONOLOGY

Chapter 17

Earth: Planet of Wonders

Let's explore Earth, the third rock from the sun and our home planet. While it seems we're on the brink of manned missions to other planets, as of writing this series, every human has lived, born, and died on Earth, with no chance to set foot elsewhere. The space age and potential colonization efforts will be discussed later. For now, let's delve into Earth's formation and its moon. Unlike Mercury and Venus, Earth has no natural satellites.

When considering the solar system's formation, we imagine a protoplanetary disk filled with heavy elements ejected into the interstellar medium during star deaths. Over time, matter in this disk coalesced into pebble-sized, then boulder-sized chunks, eventually forming large planetesimals due to gravity. These combined to form rocky objects, including the early Earth, largely molten due to immense collision-generated heat.

As most matter within Earth's orbital radius collected, collisions became rarer, allowing

Earth to cool and develop a crust. The last major collisions produced craters and released gas trapped in the rock, contributing to the early atmosphere. Comets later delivered more atmosphere and water during the late heavy bombardment.

The Earth's composition, better understood than any other solar system object, consists mainly of an iron core with some nickel, surrounded by a rocky mantle rich in silicon, aluminum, and oxygen, and a thin crust on the surface. Density determines the distribution of matter, with heavier elements at the core and lighter compounds towards the crust.

The outer core is liquid, with the mantle capable of gradual flow, causing sections of the crust to drift over time, known as continental drift. The circulation in the molten core, combined with Earth's rotation, generates an incredible magnetic field shielding Earth from cosmic rays, forming phenomena like the Northern Lights.

Moving on to Earth's moon, it resulted from a collision between two planetesimals in Earth's early stages. One became Earth, and the other, likely a Mars-sized object, collided, splashing enormous amounts of rock into space. Gravitational forces collected this debris to form the moon.

The moon, rich in silicates and poor in iron, cooled quickly due to its small size, resulting in a rocky crust covered in craters and dark patches called maria. With no atmosphere, the moon's features remain largely unchanged since its formation.

The moon orbits Earth every twenty-seven days, with its rotational period matching its orbital period, causing synchronous rotation. Tidal forces between Earth and the moon, responsible for Earth's tides, influence coastal areas, affecting high and low tides.

Tides may have played a crucial role in early life development, indicating life might not exist on Earth without the moon. From the birth of the universe to the development of life on Earth, understanding these details empowers us to explore further, perhaps venturing to Mars, the next planet humans will set foot on.

THE COSMIC CHRONOLOGY

Chapter 18
The Red Neighbour: Mars

Mars, the fourth planet from the sun, has long held a place of intrigue in human history. Its distinct reddish hue, caused by iron oxide on its surface, earned it the name "Mars," after the Roman god of war. Despite being our next-door neighbor in the solar system, Mars has always seemed both familiar and mysterious, inviting speculation about its potential as a habitat for life.

In many ways, Mars bears a resemblance to Earth. Its surface features, captured in images from rovers and orbiters, often evoke comparisons to terrestrial landscapes. From vast plains to towering mountains, Mars presents a varied and rugged terrain that hints at geological processes similar to those shaping our own planet. But beneath the surface similarities lie significant differences that make Mars a world unto itself.

One striking dissimilarity is Mars' size. While it shares similarities in some aspects, such as having a day length and seasons comparable to Earth's, Mars is significantly smaller, with about half the diameter and only one-tenth the mass of our home planet. This size difference has profound implications for Mars' geology, atmosphere, and potential for sustaining life.

Early observations of Mars fueled speculation about the possibility of life on the red planet. Italian astronomer Giovanni Schiaparelli's descriptions of "channels" on Mars, later mistranslated as "canals," sparked imaginations and led to visions of a civilization capable of engineering vast waterways across the Martian surface. These fanciful ideas of Martian civilizations, popularized in science fiction literature, persisted until clearer observations revealed a stark reality.

Modern exploration has shown that Mars is a harsh and inhospitable world. Its thin atmosphere, composed primarily of carbon dioxide with traces of nitrogen and argon, provides little protection from the harsh radiation and extreme temperatures on the surface. While Mars experiences temperatures that can be relatively mild during the day, plunging to frigid lows at night, its thin atmosphere cannot retain heat effectively, resulting in dramatic temperature fluctuations.

Surface water on Mars is scarce, existing mainly in the form of ice in polar caps and subsurface reservoirs. Evidence from orbiters and rovers suggests that Mars once had liquid water flowing on its surface, carving channels, valleys, and possibly even lakes and oceans. However, over billions of years, Mars underwent significant changes, leading to the loss of its surface water and the transformation of its climate.

The absence of a global magnetic field on Mars is another key difference from Earth. Earth's magnetic field, generated by the motion of molten iron in its outer core, serves as a protective shield against solar wind and cosmic radiation. However, Mars' smaller size and cooler interior likely led to the solidification of its core, resulting in the loss of its magnetic field billions of years ago. This loss exposed Mars' atmosphere to erosion by solar wind, further contributing to its current thin atmosphere.

Despite its harsh conditions, Mars continues to fascinate scientists and explorers alike. Robotic missions, such as NASA's Perseverance rover, continue to uncover new insights into the planet's geology, climate history, and potential for past or present life. Future human missions to Mars, while still in the realm of exploration and speculation, hold the promise of further unraveling the mysteries of the red planet and perhaps one day establishing a human presence on its surface.

As we continue to study Mars and unravel its secrets, we gain valuable insights into planetary evolution, the potential for life beyond Earth, and the challenges of human exploration of other worlds. Mars serves as both a tantalizing destination for exploration and a cautionary tale of the delicate balance that sustains life on our own planet.

In conclusion, Mars stands as a testament to the complexities of planetary science and the enduring human spirit of exploration. Its similarities to Earth, combined with its unique features and challenges, make it a compelling subject of study for scientists and a source of inspiration for future generations of explorers. Whether as a potential future home for humanity or a window into the mysteries of the cosmos, Mars continues to capture our imagination and drive our quest for knowledge

Chapter 19
The Dictator and the Saviour: Jupiter

Diving into the vast expanse of our solar system, we leave behind the terrestrial realms to venture into the domain of the gas giants, the colossal behemoths that reign supreme in the celestial tapestry. These titanic worlds, born in the frigid depths of space far from the warming embrace of the sun, owe their immense size and grandeur to the abundance of icy materials and volatile gases that pervade the outer reaches of our cosmic neighborhood. Unlike their rocky counterparts, which are predominantly composed of dense metals and silicates, the gas giants are characterized by their massive atmospheres, primarily comprised of hydrogen and helium, with traces of methane, ammonia, and other volatile compounds.

At the forefront of this celestial pantheon stands Jupiter, the mighty king of the planets, whose imposing presence dominates the solar system. Named after the supreme deity of Roman mythology, Jupiter commands reverence and awe with its colossal dimensions and formidable gravitational influence. With a diameter more than ten times that of Earth and a mass exceeding that of all the other planets combined, Jupiter is a true colossus among celestial bodies. Its composition, characterized by swirling clouds of hydrogen and helium, conceals a solid core enveloped by layers of dense gases and liquid metallic hydrogen. This unique structure, coupled with Jupiter's rapid rotation, gives rise to a dynamic and turbulent atmosphere, marked by swirling storms and vortices that dance across its cloud tops.

Among the most prominent features of Jupiter's atmosphere is the Great Red Spot, a colossal storm system that has raged for centuries, stretching across thousands of kilometers and boasting wind speeds exceeding 400 kilometers per hour. This iconic feature, larger than our entire planet, serves as a testament to the planet's tumultuous weather patterns and dynamic atmospheric processes. In addition to the Great Red Spot, Jupiter's atmosphere is adorned with a myriad of smaller storms and atmospheric disturbances, each a testament to the planet's complex and dynamic nature.

Beneath its turbulent exterior, Jupiter harbors a wealth of mysteries waiting to be unraveled. At its core lies a solid inner region comprised of iron, rock, and other heavy elements, surrounded by a vast layer of liquid metallic hydrogen. This metallic hydrogen layer, generated by the immense pressure exerted by Jupiter's gravity, is responsible for the planet's powerful magnetic field, the strongest in the solar system. This magnetic field extends far into space, creating a protective magnetosphere that shields the planet from the solar wind and cosmic radiation.

In addition to its magnetic field, Jupiter is also known for its system of faint rings, composed of fine dust particles that orbit the planet in a thin disk. Though not as prominent as Saturn's majestic ring system, Jupiter's rings serve as a reminder of the planet's gravitational influence and dynamic environment.

Yet, perhaps the most fascinating aspect of Jupiter's realm lies in its extensive family of moons, each a world unto itself with its own unique characteristics and mysteries. Among the most well-known of these moons are the Galilean moons—Io, Europa, Ganymede, and Callisto—discovered by the Italian astronomer Galileo Galilei in 1610. These moons, named after characters from Greek mythology, offer a wealth of scientific insights into the dynamics and evolution of Jupiter's system.

Ganymede, the largest moon in the solar system, boasts a diverse and geologically complex terrain, including cratered highlands, grooved terrain, and icy plains. Its surface features suggest a complex geological history, with evidence of past tectonic activity and resurfacing events. Beneath its icy exterior lies a subsurface ocean of liquid water, making Ganymede one of the most promising targets for future exploration missions.

Callisto, by contrast, is a heavily cratered world, with a surface scarred by impacts from countless asteroids and comets over billions of years. This ancient and battered moon serves as a testament to the violent history of the solar system, preserving a record of impacts dating back to the dawn of the solar system.

Io, the innermost of the Galilean moons, is a world of fire and fury, with hundreds of active volcanoes erupting across its surface. This volcanic activity is driven by tidal forces exerted by Jupiter's immense gravity, which generate intense heat within Io's interior. As a result, Io is the most volcanically active body in the solar system, with lava flows and volcanic plumes erupting hundreds of kilometers into space.

Europa, the smallest of the Galilean moons, is perhaps the most intriguing of all, with its smooth icy surface and enigmatic reddish-brown cracks. These cracks, known as lineae, are thought to be the result of tidal forces acting on Europa's icy crust, causing it to stretch and fracture. Beneath its icy exterior lies a vast ocean of liquid water, kept warm by tidal heating generated by Jupiter's gravity. This subsurface ocean, which may be in contact with Europa's rocky mantle, could harbor conditions suitable for life as we know it, making Europa a tantalizing target for future astrobiological exploration.

In summary, Jupiter and its moons present a captivating tableau of cosmic wonders, from the swirling storms of its turbulent atmosphere to the icy landscapes of its diverse moons. As we continue to unravel the mysteries of our solar system, Jupiter stands as a testament to the dynamic and ever-changing nature of our cosmic neighborhood, offering tantalizing clues to the origins and evolution of our celestial home.

Fun fact: Jupiter's huge gravitational field deflects comets and asteroids away from our delicate, rocky home planet.

Source: article by The University of California, Riverside. Published on: November of 2023

Chapter 20
Cosmic Elegance: Saturn

As we venture further into the outer reaches of our solar system, we encounter Saturn, the majestic sixth planet from the sun. Named after the Roman god of agriculture, Saturn, also known as Cronus in Greek mythology, holds a place of reverence among the celestial pantheon. Like its counterpart Jupiter, Saturn resides at a considerable distance from the sun, nearly ten astronomical units away, casting a faint glow in the distant reaches of the solar system.

Despite its remote location, Saturn commands attention with its imposing presence and enigmatic allure. With a diameter rivaling that of Jupiter and a density lower than water, Saturn defies conventional expectations, hinting at its ethereal composition. Comprised predominantly of hydrogen and helium gases, interspersed with traces of heavier elements, Saturn presents a celestial enigma veiled beneath layers of swirling clouds and gas.

Beneath its turbulent exterior lies a realm of intricate complexity, characterized by layers of atmospheric gases and exotic compounds. Helium droplets, suspended in a vast sea of hydrogen, envelop a solid metallic core, nestled at the heart of the planet. This dense core, composed of iron, rock, and water, serves as the anchor for Saturn's ethereal form, anchoring it amidst the cosmic dance of the planets.

Amidst the haze of Saturn's atmosphere, clouds of ammonia gas swirl, obscuring the planet's surface and lending it an aura of serene beauty. Unlike its tumultuous sibling Jupiter, Saturn appears tranquil and serene, its smooth surface belying the turbulent forces that churn beneath.

Yet, Saturn's most iconic feature lies in its resplendent rings, a celestial masterpiece unrivaled in the annals of astronomy. Stretching outwards from the planet's equator, these luminous bands of ice and dust captivate the imagination, evoking wonder and awe in all who behold them. Formed from the remnants of primordial gas and dust, the rings trace the ancient history of Saturn, bearing witness to the tumultuous birth of our solar system.

Despite their ethereal beauty, the rings of Saturn are shrouded in mystery, their origins and evolution the subject of ongoing scientific inquiry. While some theories posit that the rings are relics of Saturn's formation, others suggest a more recent origin, driven by the gravitational interactions of moons and debris. Regardless of their genesis, the rings of Saturn stand as a testament to the cosmic forces that shape our celestial neighborhood, their delicate structure a testament to the intricate dance of gravity and inertia.

In addition to its stunning rings, Saturn boasts a diverse array of moons, each a world unto itself, with its own unique characteristics and mysteries. Among these celestial companions, Titan reigns supreme, a world of ice and rock shrouded in an atmosphere of nitrogen and hydrocarbons. As the second-largest moon in the solar system, Titan holds the promise of extraterrestrial exploration, its frigid surface concealing untold secrets of cosmic significance.

Yet, Saturn's moons hold more than just scientific intrigue; they offer tantalizing glimpses into the potential for life beyond Earth. Enceladus, a small moon with a turbulent history, exhibits geysers of water and organic material, hinting at the presence of a subsurface ocean teeming with the building blocks of life. Likewise, Europa, akin to its Jovian counterpart, harbors a hidden ocean beneath its icy crust, offering a beacon of hope in the search for extraterrestrial life.

In conclusion, Saturn and its entourage of moons and rings present a celestial spectacle unmatched in the cosmos. From the ethereal beauty of its luminous rings to the enigmatic allure of its diverse moons, Saturn beckons us to explore the mysteries of our solar system and beyond. As we continue our journey of discovery, Saturn stands as a testament to the boundless wonders of the universe, inviting us to embark on a voyage of cosmic exploration and enlightenment.

THE COSMIC CHRONOLOGY

Chapter 21

Blue Giant: Uranus

As we journey through the solar system, our distance from the sun increases significantly with each planet we encounter. Jupiter sits at approximately five astronomical units from the sun, while Saturn lies nearly ten. Moving farther out, Uranus, the seventh planet, is positioned at nearly twenty astronomical units away from our star.

Named after the primal god of the sky, Uranus follows the mythological pattern established by its predecessors, with Cronus, or Saturn, being the father of Jupiter, or Zeus. This trend continued upon its discovery in the 18th century. Despite its smaller size compared to Jupiter and Saturn, Uranus remains substantial, boasting a diameter approximately four times that of Earth's. Its relatively diminutive size and vast distance from Earth explain why it remained unseen until the advent of telescopes.

Uranus, like its gas giant counterparts, possesses a hydrogen-rich atmosphere, with a notable presence of methane contributing to its distinct blue hue. Unlike the larger gas giants, Uranus lacks metallic hydrogen in its interior. Instead, it likely consists of hydrogen gas enveloping layers of water mixed with methane and ammonia, surrounding a core of iron and rock.

Similar to Saturn, Uranus boasts a system of rings, albeit less complex and darker in appearance. Unlike Saturn's icy rings, Uranus's rings are believed to be primarily composed of organic material. Additionally, Uranus is orbited by an extensive array of moons, numbering 27 in total. Five of these moons are sizable and bear names inspired by literary characters from the works of William Shakespeare and Alexander Pope, including Miranda, Ariel, Umbriel, Titania, and Oberon. These moons, like those of other gas giants, consist of a mixture of ice and rock.

Miranda, the smallest of Uranus's major moons, displays a peculiar patchwork surface, suggesting a violent collision between two separate bodies that merged to form the moon. This event left behind massive cliffs towering twice the height of Mount Everest.

One of Uranus's most distinctive features is its extreme tilt, with its equatorial plane nearly perpendicular to its orbit. This unique orientation results in one hemisphere experiencing prolonged daylight while the other remains in perpetual darkness for half of Uranus's year, until they swap positions during the other half. This extreme tilt is believed to be the result of a collision with a planetesimal during Uranus's early formation, an event that also contributed to the formation of its moons.

THE COSMIC CHRONOLOGY

Chapter 22
The Last Knight: Neptune

Neptune, standing as the culminating gas giant and the ultimate celestial entity within our solar system, occupies a profound distance from the radiant embrace of the sun, measuring an expansive 30 astronomical units from its luminous source. This spatial expanse mirrors the vast chasm that separates Uranus from its predecessor, Saturn. The striking resemblance in size and mass between Neptune and Uranus underscores their shared lineage, with Neptune's ethereal, cerulean complexion earning it a fitting association with the Roman deity ruling over the boundless depths of the sea.

In parallel to its cosmic kin, Neptune envelops itself in an atmospheric ensemble richly endowed with hydrogen, veiling layers of water, methane, and ammonia, ensconcing a core forged from the immutable union of iron and rock. Distinguished by its distinctive cloud bands evocative of the intricate tapestry adorning Jupiter's countenance, Neptune's atmospheric dynamics unveil transient marvels, epitomized by the ephemeral spectacle of a great dark spot, analogous to its Jovian counterpart, the renowned great red spot. These atmospheric phenomena, driven by convective currents ignited by the incandescent heart of the planet, animate Neptune's atmosphere with tempestuous winds, whose frenzied dance attains velocities unmatched anywhere within the solar expanse, cresting at an astonishing 2200 kilometers per hour.

Complementing its atmospheric splendor, Neptune's celestial adornments manifest in the form of narrow, enigmatic rings, echoing the celestial debris born of titanic collisions between diminutive moons or celestial wanderers. Laden with a profusion of cosmic dust, these enigmatic rings enshroud Neptune in an aura of mystique, weaving a narrative of celestial upheaval and cosmic ballet.

The retinue of Neptune, comprising a diverse assembly of fourteen moons, draws inspiration from the pantheon of Greek mythology, with each celestial satellite bearing the name of a minor water deity. Foremost among these celestial denizens stands Triton, a titan among moons, whose enigmatic orbit defies convention, tracing a path antithetical to Neptune's rotation in a celestial ballet reminiscent of cosmic intrigue. Triton's retrograde orbit hints at an origin distinct from Neptune's progeny, suggesting a fortuitous encounter that ensnared this celestial wanderer within Neptune's gravitational embrace. The cataclysmic event of Triton's capture likely disturbed the tranquil harmony of Neptune's celestial court, precipitating irregularities exemplified by the elliptical trajectory of Nereid, whose celestial journey traverses the furthest reaches of Neptune's gravitational influence.

Triton, akin to its celestial counterpart, Titan, boasts an atmospheric ensemble primarily composed of nitrogen, interwoven with traces of methane and carbon monoxide, veiling its frigid landscape in an ethereal shroud. Among Neptune's retinue, seven moons exhibit a prograde motion aligned with the equatorial plane, embodying the archetype of regular satellites, while the remaining seven, including Triton and Nereid, traverse irregular paths, testament to their fortuitous capture within Neptune's gravitational dominion.

The enigmatic origins of Triton beckon exploration beyond the planetary confines, inviting inquiry into the celestial tapestry woven between and beyond the planetary realms. Let us embark upon a voyage of cosmic discovery, delving into the mysteries that lie nestled within the cosmic expanse, beyond the purview of planetary realms, and unravel the enigmatic fabric of the solar system.

Chapter 23

Guardians and Separators: Asteroid Belt Kuiper Belt and Beyond

Let's conclude our exploration of the solar system, having traversed the breadth of its celestial wonders. Our odyssey commenced with an inquiry into the terrestrial quartet comprising Mercury, Venus, Earth, and Mars, before embarking on a celestial pilgrimage through the realms of the gas giants—Jupiter, Saturn, Uranus, and Neptune. Yet, our cosmic journey is far from complete, for beyond the familiar confines of the planetary realms lies a vast expanse teeming with a myriad of celestial inhabitants awaiting our scrutiny. Having plumbed the celestial depths of the terrestrial realm, our celestial odyssey propelled us beyond the familiar confines of the inner sanctum, venturing into the ethereal domains of the gas giants—Jupiter, Saturn, Uranus, and Neptune—celestial monarchs reigning supreme over the celestial hinterlands. In the regal splendour of Jupiter's celestial court, we beheld the cosmic dance of its Galilean moons, celestial pearls adorning the celestial diadem of the celestial sovereign. Saturn, with its resplendent rings, stood as a celestial sentinel guarding the celestial gates to the outer realms, its celestial allure captivating the celestial wanderer with celestial wonder. Uranus and Neptune, enigmatic siblings amidst the celestial pantheon, cast their celestial gaze upon the celestial void, their celestial presence a testament to the celestial mysteries that shroud the celestial expanse. From their celestial confines, we glimpsed the celestial ballet of moons and rings, celestial vestiges of celestial genesis echoing through the celestial corridors of time. Yet, our celestial journey is far from concluded, for beyond the planetary demesnes lie celestial frontiers teeming with celestial marvels awaiting our celestial scrutiny. Embarking on an expansive odyssey through the cosmic tapestry that adorns our solar system, we delve into two extraordinary regions: the Asteroid Belt and the Kuiper Belt. These celestial domains, shrouded in mystery and cosmic wonder, invite us to unravel their intricacies and explore the cosmic treasures they hold. Our cosmic voyage commences amidst the rocky expanse of the Asteroid Belt, a vast celestial arena stretching between the orbits of Mars and Jupiter. Here, amidst the debris of ancient cosmic collisions, a myriad of celestial wanderers drift in perpetual motion.

Navigating the celestial labyrinth of the Asteroid Belt, we encounter a diverse array of celestial denizens, each bearing witness to the tumultuous events that shaped their realm. From colossal celestial giants to diminutive celestial fragments, the Asteroid Belt presents a mosaic of celestial diversity, reflecting the dynamic history of our solar system.

At the heart of the Asteroid Belt lies Ceres, a celestial behemoth and the largest inhabitant of this cosmic realm. Once hailed as a celestial luminary in its own right, Ceres now stands as a celestial sentinel amidst the celestial expanse, its presence a testament to the celestial forces that shaped its domain.

Beyond Ceres, myriad smaller asteroids populate the Asteroid Belt, each with its own unique composition and celestial history. Some bear the scars of cosmic collisions, their rugged surfaces a testament to the violent events that shaped their celestial journey. Others harbor precious metals and minerals, remnants of the cosmic crucible from which they emerged.

Throughout the ages, the Asteroid Belt has served as a celestial laboratory, offering valuable insights into the formation and evolution of our solar system. By studying the composition and dynamics of its inhabitants, astronomers gain a deeper understanding of the celestial processes that govern our cosmic neighborhood. Venturing beyond the confines of the inner solar system, we arrive at the enigmatic realm of the Kuiper Belt—a vast expanse of icy bodies that extends beyond the orbit of Neptune. Here, amidst the frozen depths of space, celestial wanderers drift in silent orbit, their icy surfaces illuminated by the distant light of the sun.

The Kuiper Belt is home to a diverse array of celestial bodies, each with its own story to tell. From icy dwarfs to rocky remnants, the Kuiper Belt is a celestial wilderness teeming with cosmic wonders awaiting discovery.

Among the celestial denizens of the Kuiper Belt, Charon stands out as a celestial companion to Pluto, its presence a testament to the celestial forces that shaped their shared domain. Together, they form a celestial duo that embodies the celestial mysteries of the outer solar system.

Beyond the Kuiper Belt lies the scattered disk—a region of cosmic debris that extends into the outer reaches of the solar system. Here, celestial wanderers like Eris drift in silent orbit, their presence a reminder of the celestial diversity that pervades our cosmic neighborhood.

In the depths of the Kuiper Belt, comets emerge as celestial heralds of cosmic revelation, their icy surfaces tracing celestial trajectories that carry them across the celestial expanse. By studying these celestial wanderers, astronomers gain valuable insights into the celestial processes that shape our solar system. Moving further, we have The Oort Cloud and interstellar space is both located in the vast expanse of outer space, but they occupy different regions and serve distinct roles in the cosmic landscape.

The Oort Cloud is a theoretical region of space located far beyond the outermost planets of our solar system, extending to distances of up to 100,000 astronomical units (AU) from the Sun. It is thought to be a spherical shell of icy bodies, remnants from the formation of the solar system, and is believed to contain billions, if not trillions, of comets and icy planetesimals. The Oort cloud is divided into two distinct regions: the outer Oort Cloud, which extends from about 2,000 AU to 20,000 AU, and the inner Oort cloud, which extends from about 20,000 AU to 100,000 AU. This vast reservoir of icy debris is thought to be the source of long-period comets that occasionally visit the inner solar system.

On the other hand, interstellar space refers to the space between stars within a galaxy. It is the vast, empty expanse that lies beyond the influence of any individual star's gravity. Interstellar space is characterized by extremely low densities of matter, primarily composed of gas and dust, but it also contains cosmic rays, magnetic fields, and other forms of energy. Interstellar space extends throughout the entire galaxy, encompassing vast distances between stars and their associated planetary systems.

In summary, the Oort Cloud is a specific region within our solar system, while interstellar space refers to the space between stars within a galaxy, including the space beyond the outer reaches of the solar system.

As we traverse the cosmic wilderness of the Asteroid Belt and the Kuiper Belt, we are reminded of the vastness and diversity of our solar system. Each celestial body tells a story—a tale of cosmic evolution and celestial wonder that continues to unfold with each passing moment.

From the rocky expanse of the Asteroid Belt to the icy wilderness of the Kuiper Belt, our celestial journey offers a glimpse into the celestial forces that govern our cosmic domain. As we continue to explore these celestial frontiers, we uncover new mysteries and unlock the secrets of our cosmic origins.

In the cosmic symphony of celestial evolution, the Asteroid Belt and the Kuiper Belt stand as celestial witnesses to the celestial drama that has unfolded over billions of years. By studying these celestial realms, we gain a deeper understanding of the celestial processes that shape our solar system and the universe beyond.

Chapter 24
Dwarfs of the Solar System

At the heart of the Kuiper Belt lies Pluto, a celestial enigma once hailed as the ninth planet of our solar system. Wait, didn't our schools teach us that Pluto is a dwarf planet? Exactly, for decades Pluto was considered as a planet instead of calling it a Dwarf. Because, there was not specific definition called Dwarf Planet until 2006. But hold on, what is the difference between a Planet and a Dwarf planet. To know about that we must zoom out to the very beginning of the discovery. In the vast expanse of our solar system lies a small, enigmatic world that has captured the imagination of astronomers and space enthusiasts alike. This distant celestial body, known as Pluto, holds a unique place in our understanding of the outer reaches of our cosmic neighborhood. Let us embark on a journey to uncover the story behind the discovery of Pluto and the remarkable individuals who unveiled its secrets.

The tale of Pluto begins in the early 20th century, a time when astronomers were avidly searching for a ninth planet beyond the known orbits of Neptune and Uranus. Observations of these outer planets revealed discrepancies in their predicted positions, leading astronomers to hypothesize the existence of another celestial body exerting gravitational influence on their orbits.

The quest to find this elusive ninth planet gained momentum in the 1920s when astronomer Percival Lowell initiated a systematic search at the Lowell Observatory in Flagstaff, Arizona. Lowell had long been intrigued by the possibility of a trans-Neptunian planet, and he dedicated years of his life to scouring the night sky in pursuit of this celestial enigma.

Despite Lowell's diligent efforts, he did not live to witness the discovery of Pluto. However, his legacy lived on through his observatory and the dedicated astronomers who carried on his quest. One such individual was Clyde Tombaugh, a young astronomer from Kansas who joined the Lowell Observatory in 1929 as a staff member.

Under the guidance of the observatory's director, Vesto Melvin Slipher, Tombaugh embarked on the monumental task of systematically photographing and analyzing vast swathes of the night sky. Using a device called a blink comparator, Tombaugh meticulously compared pairs of photographic plates taken days apart to detect any celestial objects that appeared to move against the backdrop of stars.

On February 18, 1930, Tombaugh made a historic discovery that would change our understanding of the solar system forever. In examining photographic plates taken in January of that year, Tombaugh noticed a faint dot of light that shifted its position slightly over successive nights. This celestial wanderer, later designated Pluto, was the elusive ninth planet that astronomers had been searching for.

The announcement of Pluto's discovery sent shockwaves through the astronomical community and captured the imagination of the public worldwide. Named after the Roman god of the underworld, Pluto became a symbol of the mysterious depths of the outer solar system. It was all going good, as technology advanced and our understanding of the outer solar system deepened, Pluto's status as a planet came into question. In 2006, the International Astronomical Union redefined the criteria for what constitutes a planet, leading to Pluto's reclassification as a dwarf planet. Now, Pluto was lying in the Kuiper belt, it had few moons too, but, when observations proved that the Kuiper belt had thousands of same sized planetesimals. IAU (International Astronomical Union) then had come to a conclusion that instead of naming thousands as planets, we must refine the definition of Planet. Then after there were few conditions for qualifying as a Planet.

The object must be big enough that its gravity clears its orbit, it must be orbiting a star and it must be big enough that its gravity must shape it a sphere.

Unfortunately, Pluto could not pass all those conditions. So therefore, Pluto's status as a planet came into question. In 2006, the International Astronomical Union redefined the criteria for what constitutes a planet, leading to Pluto's reclassification as a dwarf planet. Despite this change in designation, Pluto remains a fascinating and scientifically valuable world. In 2015, NASA's New Horizons spacecraft conducted a historic flyby of Pluto, providing humanity with the first up-close images and scientific data about this distant world.

The discovery of Pluto stands as a testament to the power of human curiosity and the tireless dedication of astronomers to unravel the mysteries of the cosmos. From Percival Lowell's visionary quest to Clyde Tombaugh's meticulous observations, the story of Pluto's discovery is a testament to the indomitable spirit of exploration that drives humanity's quest to understand the universe.

Not just Pluto but, we have 5 Dwarf planets in our solar system. Let's go through them one by one.

Ceres:

Ceres, the largest object in the asteroid belt, occupies a pivotal position in our understanding of the solar system's formation and evolution. Discovered by Giuseppe Piazzi in 1801, it initially sparked debate about its classification, oscillating between planet, asteroid, and finally, dwarf planet status.

With a diameter of approximately 940 kilometers, Ceres stands as a testament to the diversity of celestial bodies populating our cosmic neighborhood. Its composition, predominantly rock and ice, hints at its role as a remnant from the early solar system's tumultuous past.

In 2015, NASA's Dawn spacecraft embarked on a historic mission to orbit Ceres, offering unprecedented insights into its surface features. Dawn's observations unveiled a complex world characterized by impact craters, mountain ranges, and mysterious bright spots within craters, possibly indicative of subsurface water ice.

Ceres' unique composition and proximity to Earth make it an enticing target for future exploration. Scientists hope to unravel the secrets hidden beneath its icy surface, shedding light on the processes that shaped our solar system billions of years ago.

Haumea:

Haumea, a distant resident of the outer solar system, challenges our perceptions of celestial bodies with its unusual elongated shape and rapid rotation. Discovered in 2004 by Mike Brown and his team, Haumea stands out as a celestial oddity among the diverse population of dwarf planets.

Named after the Hawaiian goddess of childbirth, Haumea's elongated shape suggests a tumultuous past marked by violent collisions and gravitational interactions. Its flattened ellipsoid form sets it apart from its spherical counterparts, inviting speculation about its origins and evolution.

Haumea's surface, covered in crystalline water ice, reflects sunlight to create a bright, reflective appearance. Its unique characteristics offer valuable insights into the dynamic processes shaping the outer solar system and the diversity of celestial bodies populating its remote realms.

Makemake:

Makemake, a distant denizen of the Kuiper belt, beckons astronomers with its enigmatic presence and reddish-brown complexion. Named after the creator god of Easter Island, Makemake was discovered in 2005 by Mike Brown's team at the Palomar Observatory.

With a diameter of approximately 1,430 kilometers, Makemake ranks among the largest dwarf planets in the solar system. Its surface, coated in frozen methane, contributes to its distinctive coloration and reflects sunlight across the vast expanse of the Kuiper belt.

Makemake's orbit, inclined relative to the plane of the solar system, hints at its turbulent history and gravitational interactions with neighboring celestial bodies. Its remote location and unique properties offer a glimpse into the outer reaches of our cosmic neighborhood and the processes shaping its evolution.

Eris:

Eris, a distant dweller of the Kuiper belt, captivates astronomers with its massive size and eccentric orbit. Discovered in 2005 by Mike Brown's team, Eris played a pivotal role in redefining our understanding of dwarf planets and the dynamic nature of the solar system.

With a diameter of approximately 2,326 kilometers, Eris ranks among the largest dwarf planets, challenging Pluto's status as the ninth planet. Its surface, coated in nitrogen ice, reflects sunlight across the frigid expanse of the Kuiper belt, offering a glimpse into the outer reaches of our cosmic neighborhood.

Eris' eccentric orbit, spanning from 38 to 97 astronomical units from the Sun, hints at its turbulent past and gravitational interactions with neighboring celestial bodies. Its discovery has reshaped our understanding of the outer solar system and the diverse array of celestial bodies populating its remote realms.

Chapter 25
An Unexpected Miracle

In the grand narrative of our planet, life's arrival is a mere blink of an eye. Earth, estimated to be around 4.5 billion years old, spent its initial years in a chaotic dance of fire and fury. The young Earth was a molten hellscape, bombarded by celestial debris. As it gradually cooled and solidified, a thick atmosphere choked the planet, dominated by noxious gases like methane, ammonia, and carbon dioxide – a far cry from the life-sustaining haven it would become.

The embers of life, if they existed then, wouldn't have left any recognizable fossils. But scientists believe the stage was being set for a grand transformation. Powerful bolts of lightning ripping through the atmosphere, along with intense UV radiation from the young sun, might have played a crucial role.

These energetic forces could have kickstarted a symphony of chemical reactions within this primordial soup, giving rise to the building blocks of life – organic molecules like amino acids and nucleotides.

One prevailing theory, known as the Miller-Urey experiment, demonstrates this possibility. In a simulated early Earth environment, scientists provided the necessary ingredients – water, methane, ammonia, and hydrogen – and bombarded them with energy. The result? The formation of simple organic molecules, hinting at the potential for life's pre-requisites to arise naturally.

Another theory suggests clay minerals might have played a role. These microscopic structures could have acted as tiny cradles, attracting and concentrating organic molecules, facilitating their interaction and potentially even replication. These early molecules, lacking the complex structures of living organisms, may have begun to self-assemble into even more intricate configurations, driven by the laws of physics and chemistry.

The next critical step remains shrouded in mystery. How did these simple molecules transform into something we can call alive? Theories abound. One possibility involves the formation of self-replicating molecules like RNA, which can both store information and act as a rudimentary enzyme. The ability to replicate would have been a game-changer, allowing these early entities to create copies of themselves and potentially evolve over time.

Another theory proposes the emergence of proto-cells – enclosed compartments formed by fatty acid-like molecules (lipids). These compartments could have harbored the necessary ingredients for life, separating them from the harsh external environment. Over time, these proto-cells might have become more sophisticated, developing membranes to regulate what entered and exited, and evolving internal machinery for rudimentary metabolism.

The timeline for these events is blurry. Estimates suggest life emerged sometime between 4.3 and 3.8 billion years ago. The oldest fossils we have discovered, microbial structures called stromatolites, date back to 3.5 billion years. These tiny rock-like formations, created by layered colonies of microbes, are a testament to the tenacity of life in its nascent form.

The early inhabitants of Earth were likely single-celled organisms, far simpler than anything we see today. They probably didn't have the complex internal structures or the oxygen-dependent respiration that characterizes most modern life. They may have relied on simpler energy-harvesting mechanisms, perhaps harnessing geothermal vents or the energy from sunlight in a way very different from photosynthesis.

These early life forms, however, held the spark that would ignite a firestorm of evolution. Through natural selection, random mutations that provided an advantage – a more efficient way of acquiring nutrients, perhaps, or a better defense against the harsh environment – would have been passed on to offspring. Over vast stretches of time, these incremental changes would have led to a dazzling diversification of life.

The emergence of photosynthesis, estimated to have occurred around 2.4 billion years ago, marked another pivotal moment. Cyanobacteria, a type of bacteria, began harnessing sunlight to convert water and carbon dioxide into organic molecules and releasing oxygen as a byproduct. This innovation not only revolutionized the way life utilized energy but also fundamentally altered Earth's atmosphere. Over time, oxygen began to accumulate, paving the way for more complex organisms that relied on this vital element for respiration.

The origin of life remains an active area of research, with new discoveries constantly challenging and refining our understanding. From the primordial soup to the first self-replicating molecules and the emergence of the first cells, the story of life's beginnings is a testament to the enduring power of nature's ingenuity. It is a story that continues to unfold, with each new discovery offering a deeper glimpse into the remarkable genesis of life on our planet.

In the infancy of our planet, amidst the cacophony of volcanic eruptions and meteorite impacts, the seeds of life were sown. Earth, estimated at a staggering 4.5 billion years old, was a tumultuous cauldron of molten rock and a choking atmosphere of methane, ammonia, and carbon dioxide. It was a far cry from the life-sustaining haven it would eventually become.

The nascent stages of life wouldn't have left behind readily identifiable fossils. However, scientists believe the stage was being meticulously set for a grand transformation. Powerful bolts of lightning crackling through the atmosphere, along with the intense ultraviolet radiation from the young sun, might have played a catalytic role. These energetic forces could have initiated a symphony of chemical reactions within this primordial soup, giving rise to the building blocks of life – organic molecules like amino acids, the fundamental units of proteins, and nucleotides, the building blocks of DNA and RNA.

The Miller-Urey experiment, conducted in the mid-20th century, offers a compelling illustration of this possibility. Scientists simulated the early Earth environment within a laboratory apparatus, providing water, methane, ammonia, and hydrogen gas as the key ingredients, and then bombarded the mixture with sparks to mimic lightning. The result? The formation of simple organic molecules, hinting at the potential for life's precursors to arise naturally from the raw materials present on early Earth.

Another theory proposes clay minerals as potential midwives of life. These microscopic structures could have acted as tiny cradles, attracting and concentrating organic molecules, facilitating their interaction and potentially even replication. These early molecules, lacking the complex structures of living organisms, might have begun to self-assemble into even more intricate configurations, driven by the inherent laws of physics and chemistry.

However, the exact transition from these simple molecules to what we can call "alive" remains shrouded in mystery. Theories abound, each attempting to illuminate this crucial step. One possibility involves the formation of self-replicating molecules like RNA, which can both store information like DNA and act as a rudimentary enzyme. This ability to replicate would have been a game-changer, allowing these early entities to create copies of themselves and potentially evolve over time through natural selection.

Another theory proposes the emergence of proto-cells – enclosed compartments formed by fatty acid-like molecules (lipids). These compartments could have harbored the necessary ingredients for life, separating them from the harsh external environment. Over time, these proto-cells might have become more sophisticated, developing membranes to regulate what entered and exited, and evolving internal machinery for rudimentary metabolism, the process of converting raw materials into energy.

The timeline for these events is blurry, with estimates placing the emergence of life sometime between 4.3 and 3.8 billion years ago. The oldest fossils we have discovered, microbial structures called stromatolites, date back to 3.5 billion years. These tiny rock-like formations, created by layered colonies of microbes, are a testament to the tenacity of life in its nascent form.

The early inhabitants of Earth were likely single-celled organisms, far simpler than anything we see today. They probably lacked the complex internal structures or the oxygen-dependent respiration that characterizes most modern life. They may have relied on simpler energy-harvesting mechanisms, perhaps harnessing geothermal vents or the energy from sunlight in a way very different from photosynthesis, the process that utilizes sunlight, water, and carbon dioxide to produce energy and oxygen.

These early life forms, however, held the spark that would ignite a firestorm of evolution. Through natural selection, random mutations that provided an advantage – a more efficient way of acquiring nutrients, perhaps, or a better defense against the harsh environment – would have been passed on to offspring. Over vast stretches of time, these incremental changes would have led to a dazzling diversification of life.

The emergence of photosynthesis, estimated to have occurred around 2.4 billion years ago, marked another pivotal moment. Cyanobacteria, a type of bacteria, began harnessing sunlight to convert water and carbon dioxide into organic molecules and releasing oxygen as a byproduct. This innovation not only revolutionized the way life utilized energy but also fundamentally altered Earth's atmosphere. Over time, oxygen began to accumulate, paving the way for more complex organisms that relied on this vital element for respiration.

The story of life's origin extends beyond Earth itself. The possibility of panspermia, the theory that life's building blocks originated in space and were delivered to Earth by comets or meteoroids, is an intriguing concept that continues to be explored. Additionally, the recent discovery of potentially habitable exoplanets – planets orbiting stars outside our solar system – raises the tantalizing possibility that life might exist elsewhere in the universe.

The origin of life remains an active area of research, with new discoveries constantly challenging and refining our understanding. From the primordial soup to the first self-replicating molecules and the emergence of the first cells, the story of life's beginnings is a testament to the enduring power of nature's ingenuity. It is a story that continues to unfold with each new discovery offering a deeper glimpse into the remarkable genesis of life on our planet.

The quest to understand life's origin extends far beyond Earth. Panspermia, the theory that life's building blocks originated in space and were delivered to early Earth by comets or meteoroids, is a fascinating concept that continues to be explored. Scientists hypothesize that these celestial wanderers could have carried organic molecules or even primitive life forms in their icy interiors, seeding life on our planet. While there's no direct evidence to confirm panspermia, it remains a viable hypothesis, especially considering the prevalence of organic molecules found in meteorites and comets.

The recent discovery of exoplanets – planets orbiting stars outside our solar system – adds another layer of intrigue to the story. With thousands of exoplanets now identified, and some residing within the habitable zone of their stars – the region where liquid water, a key ingredient for life as we know it, could exist on the planetary surface – the possibility of life arising elsewhere in the universe becomes increasingly plausible.

Future missions designed to study the atmospheres and surfaces of exoplanets will play a crucial role in unraveling this mystery. Telescopes with ever-increasing sensitivity might allow us to detect biosignatures – potential signs of life – in the atmospheres of these distant worlds. The search for extraterrestrial life is no longer science fiction; it's a scientific endeavor with the potential to rewrite our understanding of life's origins and its ubiquity in the cosmos.

Back on Earth, scientists continue to explore various avenues in the search for the spark of life. Research into the potential role of hydrothermal vents on the ocean floor offers valuable insights. These deep-sea vents spew out superheated, mineral-rich water that could have provided the perfect environment for the formation of complex organic molecules and the emergence of early life forms. Additionally, research into the self-assembly properties of certain molecules and the potential for prebiotic chemistry on other celestial bodies like Mars keeps pushing the boundaries of our knowledge.

The origin of life remains an enigma, a captivating puzzle that continues to challenge and inspire scientists across disciplines. As we delve deeper into the past, unlocking the secrets hidden within ancient rocks and extraterrestrial material, we inch closer to understanding the extraordinary events that led to the emergence of life on our planet and perhaps life elsewhere in the universe. It's a story etched not just in fossils and biosignatures, but in the very essence of our being, a reminder of our profound connection to the cosmos.

Chapter 26

The Early Conquerors: The Age of Dinosaurs

Dinosaurs, the "terrible lizards" that ruled Earth for over 180 million years, continue to spark our imaginations. Their immense size, diverse forms, and mysterious extinction leave us with a perpetual sense of wonder, urging us to understand their lost world. Let's delve into the fascinating narrative of these prehistoric giants.

Their reign wasn't an overnight takeover. Dinosaurs weren't the first reptiles to walk the Earth. Their ancestry stretches back to the Permian period, around 252 million years ago. These early reptiles diversified into a variety of forms, with some eventually evolving into the first dinosaurs during the Triassic period (243-233 million years ago).

These early dinosaurs were relatively small and agile compared to their later kin. Many were bipedal, meaning they walked on two legs. They weren't the dominant land animals yet, sharing the landscape with other reptiles and the first mammal ancestors.

The story takes a dramatic turn at the end of the Triassic period. A mass extinction event wiped out many species, creating a window of opportunity for dinosaurs to fill the ecological niches left vacant. The Jurassic period (201-145 million years ago) witnessed the rise of the iconic sauropods – the massive, long-necked herbivores like Brachiosaurus and Diplodocus. These gentle giants were the largest land animals ever to walk the Earth, their long necks allowing them to browse on the highest leaves in the lush Jurassic forests.

The Jurassic also saw the emergence of theropods, bipedal carnivores with sharp teeth and claws. These included the fearsome Allosaurus, a predator well-adapted for taking down large prey. But theropods weren't all monstrous hunters. Some, like the early ancestors of birds, were smaller and likely omnivores, feeding on a combination of insects, small animals, and fruits.

The Cretaceous period (145-66 million years ago) marked the golden age of dinosaurs. This period saw a diversification of theropods, with the rise of iconic predators like Tyrannosaurus Rex, a colossal hunter with powerful jaws and bone-crushing bite force. Maniraptorans, a group of theropods that included velociraptors and the ancestors of birds, also flourished during this time. These agile predators were known for their intelligence and pack-hunting behavior.

Herbivores diversified as well, with ceratopsians like Triceratops sporting bony frills and horns, and ankylosaurs developing thick armor for defense. Stegosaurus, with its bony plates and spiked tail, was another fascinating herbivore of the Cretaceous. The Cretaceous period also witnessed the rise of flowering plants, which may have played a role in the diversification of herbivorous dinosaurs, providing them with a more nutritious and abundant food source.

However, the reign of the dinosaurs wasn't destined to last forever. Roughly 66 million years ago, their dominance came to an abrupt end with the Cretaceous-Paleogene extinction event. A giant asteroid impact, or perhaps a series of volcanic eruptions, are thought to be the primary culprits. The event caused a dramatic global climate shift, leading to a long winter and a decline in plant life. This, in turn, disrupted the food chain, causing mass extinctions. Most dinosaurs, unable to adapt to the rapidly changing environment, perished.

Despite their extinction, dinosaurs left behind a rich fossil record that continues to be studied by paleontologists. These fossils provide us with valuable clues about their anatomy, behavior, and ecological roles. We can learn about their growth patterns by studying growth rings in fossilized bones, similar to tree rings. Footprint fossils tell us about their movement and social behavior. By studying the wear and tear on teeth, paleontologists can even make inferences about their diet.

The study of dinosaurs not only helps us understand the past but also sheds light on evolution, extinction events, and the history of life on Earth. It allows us to explore the concept of deep time – the vast stretches of time over which life has evolved. Dinosaurs remind us that even the most dominant species can be vulnerable to large-scale environmental changes. Their story serves as a cautionary tale for our own time, as we grapple with the challenges of climate change and biodiversity loss.

The legacy of dinosaurs extends far beyond the scientific realm. They continue to inspire artists, captivating audiences with their grandeur and ferocity. Movies, books, and documentaries bring these prehistoric creatures to life, sparking curiosity and wonder in generations of children. The story of dinosaurs is a reminder of the incredible diversity of life on Earth and the dramatic changes our planet has undergone. Their legacy will undoubtedly continue to inspire us for generations to come... It was not the end of life, in fact this event led to a very marvelous species evolved soon later. They are the most advanced form of life on earth. They are known as the "HUMANS".

Chapter 27
Advancement in Genes

The Cretaceous-Paleogene extinction event, a cosmic punctuation mark, slammed the curtain shut on the reign of the dinosaurs. A colossal asteroid, a celestial harbinger of doom, pummeled the Earth 66 million years ago, unleashing an inferno of unimaginable proportions. Dust choked the skies, obscuring the sun's life-giving rays. The ensuing darkness triggered a global winter, sending temperatures plummeting and plunging much of the planet into a frigid wasteland. Lush ecosystems, once teeming with dinosaurs, lay barren and silent – a stark reminder of the event's devastating impact.

However, life, in its tenacious way, possesses an incredible capacity for resilience. From the ashes of this catastrophe, a new chapter in Earth's history began to unfold – the Paleocene epoch (66 – 56 million years ago).

The immediate aftermath resembled a desolate scene from a science fiction movie. The dominant ecological players, the dinosaurs, were gone, leaving behind a vast ecological void. This opened a window of opportunity for the previously overshadowed mammals, who began a slow but remarkable diversification.

Early mammals of the Paleocene were a motley crew. Many were small, shrew-like creatures, scurrying around the undergrowth in search of insects and carrion left behind by the extinction event. However, within this unassuming group resided the seeds of future diversity. Some mammals, like the ancestors of primates and rodents, began to develop larger brains and more complex social structures. Others, like the ancestors of whales, started venturing into the aquatic realm, a testament to the versatility of mammalian evolution.

The Paleocene epoch, however, was a time of relative instability. The environment was still recovering from the cataclysm, with frequent climate fluctuations and volcanic activity. It wasn't until the Eocene epoch (56 – 33.9 million years ago) that things began to stabilize. Temperatures rose, ushering in a period of global warming. Lush forests spread across the continents, creating a verdant paradise for the burgeoning diversity of life.

This period witnessed a spectacular radiation of mammals. The ancestors of modern horses, for instance, evolved from small, dog-like creatures into browsing herbivores. Early primates, our distant ancestors, began to diversify in the Eocene rainforests. One crucial development during this time was the evolution of grasping hands and feet, an adaptation that facilitated arboreal locomotion and manipulation of objects. This dexterity, a cornerstone of human evolution, laid the groundwork for our future tool use and technological advancements.

From a paleontological perspective, the Eocene epoch is a treasure trove of fossils. Fossilized footprints tell us about the movement and social behavior of early mammals. By studying the wear and tear on teeth, paleontologists can even make inferences about their diet. These fossilized clues, like whispers from the distant past, help us piece together the fascinating story of our lineage.

However, the story of human evolution extends beyond the realm of fossils. Advances in genetics and molecular biology have given us new tools to explore our evolutionary history. By studying the DNA of living primates and comparing it to our own, scientists can build phylogenetic trees – a branching diagram that depicts the evolutionary relationships between different species. These techniques have confirmed what paleontology has long suggested – that humans share a common ancestor with chimpanzees and bonobos, our closest living relatives.

The journey from small, shrew-like mammals scurrying through the undergrowth to the large-brained, bipedal creatures we are today has been long and arduous. It's a story etched in fossils, encoded in our DNA, and ultimately, a testament to the power of natural selection. The extinction event that wiped out the dinosaurs might seem like a tragedy, but from an evolutionary perspective, it created a unique opportunity for mammals – and ultimately, for humans – to emerge and thrive.

While the Eocene epoch witnessed the diversification of many mammalian lineages, the path to modern humans still stretched far into the future. The next chapter in our evolutionary story would involve crucial developments like the rise of bipedalism, the expansion of brain size, and the development of language - all features that would ultimately define the genus Homo and its most prominent member – Homo sapiens.

This chapter marks a turning point in the grand narrative of life on Earth. From the devastation of the extinction event to the flourishing diversity of the Eocene, the stage is set for the emergence of our own lineage. The story of human evolution, with its twists and turns, is a testament to the enduring power of life and the ever-changing tapestry of the natural world. It serves as a reminder that we are not separate from, but rather a part of, this extraordinary saga. Their evolution was not the point but their advancement towards the future and the technology made the most differences. Inventions are the matter of fascination. Let's find out what was the reason to their fascination to make thing more advanced.

Chapter 28
Over the time of humans in Astronomy

Note that this chapter has skipped many human activities in survival and jump to the era where humans first started journey in Astronomy. Our story begins not ten billion years ago, but far grander. It starts with the universe's birth, the formation of stars and galaxies, culminating in the emergence of life on Earth. Here, a new chapter unfolds: the human story.

As civilizations blossomed, our ancestors turned their curious gaze skyward. Theirs was a world virtually untouched by light pollution, revealing a breathtaking expanse of stars seemingly fixed on a celestial dome. These constellations, formed by connecting starlight, were born from human imagination and a touch of wonder.

Beyond the stars, other celestial bodies captivated their attention. The Sun and Moon, along with five planets, each with its unique path across the heavens, further fueled their curiosity. These planets, named for Roman deities (names we still use today), were initially believed to revolve around an Earth-centric universe – a natural conclusion in the absence of scientific knowledge.

The path traced by these planets, known as the ecliptic, held a special significance. It was along this line that eclipses, those awe-inspiring celestial encounters, could occur. However, the true significance of the ecliptic, its connection to the plane of our solar system, remained a mystery for the time being.

Our ancestors weren't merely passive observers; they were keen observers of time. The most basic unit, the day, arose from the Sun's predictable cycle of rising and setting. Observing the Moon's phases, a rhythmic dance from full to new and back again, provided them with the concept of a month. Finally, the changing seasons, marked by periods of warmth and cold, led to the understanding of a year. These units, fundamental to our human experience, have no universal meaning but are deeply ingrained in our existence on Earth.

Subtle observations accumulated over generations. The stars, seemingly fixed, revolved around a single point in the sky – the north celestial pole for the Northern Hemisphere. This point, aligned with Earth's rotational axis, explained why the North Star remained stationary. Its significance for navigation would extend for millennia to come.

However, a closer look revealed movement. The celestial sphere, due to Earth's journey around the Sun, displayed a gradual shift throughout the year. This meant different constellations became visible at different times, a vital marker of the passage of time. It wasn't a divine message, but simply the consequence of the Sun's brilliance obscuring stars in its path.

The Sun's annual path, the ecliptic, also held a connection to the seasons. But what caused these seasonal variations? A common misconception attributed it to Earth's distance from the Sun. However, the simultaneous occurrence of summer in one hemisphere and winter in the other disproved this theory.

The answer lay not in distance, but in the tilt of Earth's rotational axis. This axis, not perpendicular to the plane of the solar system, is tilted at an angle of 23.5 degrees. This tilt creates a scenario where one hemisphere receives more direct sunlight than the other at certain times of the year. This explains why the Northern Hemisphere experiences summer when the Sun shines directly upon it, while the Southern Hemisphere endures winter's chill. This tilt also explains the varying angle of the ecliptic throughout the year.

The interplay between Earth's tilt and its orbit creates further celestial markers. The spring and fall equinoxes mark the days when the Sun crosses the celestial equator. Similarly, the summer and winter solstices denote the days when the Sun reaches its furthest points from the celestial equator, rising and setting farthest north or south, respectively. These days mark the transitions between the seasons, a knowledge that predated the understanding of the underlying cause.

Ancient monuments like Stonehenge stand as testaments to these observations. Its aligned stones frame the sunrise and sunset on specific days, showcasing a profound understanding of celestial phenomena. Similar alignments are found in pyramids and temples, not evidence of alien intervention, but the result of meticulous observation passed down through generations.

Finally, the Moon, with its ever-changing appearance, held a powerful sway over our ancestors. The lunar phases, its transition from full to new and back again, were a source of fascination. Contrary to popular belief, these phases have nothing to do with shadows cast upon the Moon. Only the Moon's sunlit face is visible from Earth, leading to the observed variations. When fully illuminated, it appears as a full moon...

Later explorations in instruments and measurements

While Aristarchus of Samos dared to propose a heliocentric model in ancient Greece, his ideas weren't widely accepted. The geocentric model, with Earth at the center, remained the dominant view for centuries.

In the 2nd century Egypt, Ptolemy perfected the geocentric model. He addressed challenging phenomena like the retrograde motion of planets, where they appear to briefly reverse course in the sky.

Retrograde motion was a major hurdle for the geocentric model. Ptolemy proposed that planets moved on epicycles – smaller circles that revolved around larger circles representing the planets' main orbits. This model could predict planetary motions with reasonable accuracy, but it wasn't perfect.

Over time, the model became increasingly complex, requiring different formulas for each planet. By the 16th century, the geocentric model felt cumbersome and in need of a major overhaul.

The time was ripe for a paradigm shift. Nicolaus Copernicus, a Polish astronomer, emerged as the champion of the heliocentric model. He meticulously studied planetary motions and demonstrated that placing the Sun at the center resolved many issues plaguing the geocentric model.

With the Sun at the center, objects closer to it orbited faster, and the retrograde motion of planets like Mars became a natural consequence of Earth's own motion in its orbit. As Earth overtakes Mars, Mars appears to move backward momentarily, similar to how a slower car appears to move backward when you pass it on the highway.

Copernicus even devised ingenious geometric calculations to estimate the distances from the Sun to each planet with remarkable precision.

However, the heliocentric model faced a challenge. Critics argued that if Earth orbited the Sun, the apparent position of stars should shift slightly throughout the year. This stellar parallax, caused by Earth's movement around the Sun, is incredibly small due to the vast distances involved. It would take telescopes far more powerful than those available in Copernicus' time to definitively detect it.

Imagine holding a finger in front of your close your eyes one at a time. Your finger appears to shift slightly because your viewpoint changes. The farther you extend your finger, the smaller this shift becomes.

Analogy can help us grasp stellar parallax. Imagine Earth's position on opposite sides of the Sun as your two eyes. Distant stars will exhibit a slight shift in their apparent position, but these distances are so immense that we need sophisticated telescopes to measure it.

By measuring this parallax shift and applying trigonometry, astronomers can calculate the distances to stars. The consistent observation of parallax strongly corroborates the heliocentric model. Over the past few centuries, advancements in telescopes and measurement techniques have solidified its validity.

The Copernican revolution wasn't just a scientific shift; it had profound cultural ramifications. The idea that Earth wasn't the center of the universe challenged long-held beliefs about humanity's place in the cosmos.

Astronomers like Giordano Bruno took it further. If the Sun is just one star among countless others, with the possibility of planets and life, then humanity wouldn't be the sole focus of creation.

This challenged the authority of the Catholic Church, which saw such thinking as a threat. Bruno was tragically tried as a heretic and burned at the stake. This dark episode serves as a reminder of the importance of free speech and access to knowledge.

We've witnessed the rise and acceptance of the heliocentric model. But the journey of astronomical discovery continues. In the next chapter, we'll explore how astronomers built upon Copernicus' work and pushed the boundaries of our understanding of the universe.

We've explored some of the celestial observations likely made by early civilizations worldwide. But science goes beyond mere observation; it's about what we do with those observations.

Science involves seeking explanations, creating models, and making predictions. We take measurements and compare them to our predictions. If they don't match, we refine the model and try again.

After centuries of observation, astronomy adopted a more mathematical approach. Some of the earliest known scientific calculations emerged during the classical period of astronomy in Ancient Greece and other contemporary civilizations.

Let's delve into some of these calculations and their revelations. First came the realization of a spherical Earth. This idea surfaced around Pythagoras' time, but initially lacked a strong logical foundation. The emphasis was on the aesthetic perfection of the sphere, making it less scientific.

However, a more logical approach arose shortly after with Aristotle. He observed that during a lunar eclipse, the Earth's shadow cast upon the Moon had a curved edge, hinting at Earth's spherical shape. Additionally, it was recognized that the visible stars depended entirely on one's location.

Traveling north or south revealed a completely new set of stars, while familiar ones disappeared. This is easily explained by a round Earth, as the other half of the celestial sphere is only visible from the other half of the globe.

Once the Earth's sphericity was established, the next logical step was to measure its dimensions. Eratosthenes achieved this with remarkable accuracy.

He reasoned that when the Sun is directly overhead at one location, it would cast no shadow on an object far enough away. He used a well in one part of Egypt and an obelisk in another for his measurements.

At noon on the summer solstice, the Sun shone directly down the well, illuminating its bottom. Simultaneously, the Sun cast a shadow on the obelisk in Alexandria, revealing that the Sun was slightly off-center.

Imagine drawing lines from the well's bottom to the Earth's center and back up to the obelisk's base. Basic geometry shows that the angle formed by these lines is seven degrees, representing a little less than 1/50th of Earth's circumference.

The distance between these locations was known to be 5,000 stadia. Using a simple ratio, we can calculate the circumference to be around 250,000 stadia, which translates to roughly 25,000 miles.

Considering the limited tools available (primarily the naked eye), these demonstrations are reproducible – if you're curious to try!

With the Earth's shape established, the next question was the distance and size of celestial objects. Incredibly, some of these quantities were also deduced.

Take the Moon, for example. Around Eratosthenes' time, another Greek named Aristarchus conducted similar work. He observed the Earth's shadow on the Moon during a lunar eclipse. By comparing the curvature of the shadow to the Moon itself, he estimated the Moon's diameter to be roughly one-third that of Earth.

He also made estimations regarding the relative distances to the Moon and Sun. While not as accurate as his other work, he was the first to propose that the Sun is much larger than Earth and even suggested the Earth revolved around the Sun.

However, there wasn't enough evidence at the time. The geocentric model**, with its rotating celestial sphere, remained dominant for centuries. Eventually, we corrected this misconception, and as this shift was a defining moment in astronomy, we'll explore it next.

For many centuries, the dominant view of the cosmos reigned supreme: the geocentric model**. This model placed Earth at the very center of the universe, with the Sun, Moon, planets, and stars all dutifully revolving around it. It seemed like a matter of common sense – after all, we stand upon a seemingly immobile Earth, while the heavens appear to majestically rotate around us in a grand celestial ballet.

However, as scientific observations grew more sophisticated and precise, cracks began to appear in the geocentric model*'s seemingly flawless facade. One of the most significant challenges arose from the observed motions of the planets. Their paths across the night sky, while seemingly following a general eastward direction, were not perfectly circular. They exhibited a peculiar phenomenon known as retrograde motion, where the planets would appear to briefly reverse their course, move westward for a short time, and then resume their eastward trek. This puzzling retrograde motion could not be easily explained by the geocentric model*, which envisioned the planets following simple, circular paths around a central Earth.

Imagine trying to explain this phenomenon using the geocentric model**. In this view, the planets were attached to crystal spheres that rotated around the Earth. These spheres were supposed to move at constant speeds, resulting in perfectly circular paths for the planets. Yet, the observed retrograde motion defied this logic. It was as if the planets were waltzing across the sky, occasionally taking a few awkward steps backward before rejoining the celestial dance.

This anomaly wasn't the only problem. The geocentric model* also struggled to explain the varying brightness of the planets. If the planets were all orbiting the Earth at relatively fixed distances, their brightness shouldn't change significantly. Yet, observations showed that planets like Mars and Venus appeared much brighter at certain points in their cycles compared to others. The geocentric model*, despite its initial appeal, simply couldn't account for these complexities. As scientific inquiry progressed, astronomers looked for alternative explanations.

One such astronomer was Aristarchus of Samos, who lived in Ancient Greece around

280 BC. He dared to propose a heliocentric model, where the Sun, not the Earth, was at the center of the universe. While his ideas weren't widely accepted at the time, they planted a seed that would continue to germinate in the minds of future astronomers.

Centuries later, during the Renaissance, another astronomer named Nicolaus Copernicus revisited the heliocentric model. Copernicus, disillusioned with the complexities of the geocentric model*, spent decades meticulously studying planetary motions. He developed a mathematical model that placed the Sun at the center of the solar system, with the Earth and other planets revolving around it in circular orbits.

Copernicus's heliocentric model wasn't without its problems. It still relied on circular orbits, which didn't perfectly align with observations. However, it offered a simpler and more elegant explanation for planetary motions, including the puzzling retrograde motion. The model also addressed the issue of varying brightness, as planets closer to the Sun would naturally appear brighter.

The heliocentric model faced significant opposition from the established scientific and religious communities. The idea of a stationary Earth challenged long-held beliefs about humanity's place in the universe. However, with the invention of the telescope in the early 17th century, astronomers like Galileo Galilei were able to gather new evidence that supported the heliocentric model. Galileo's observations of the phases of Venus, for example, were a major blow to the geocentric model*.

The heliocentric model, despite initial resistance, slowly gained acceptance throughout the 17th century. It marked a paradigm shift in our understanding of the cosmos, dethroning the Earth from its central position and paving the way for further advancements in astronomy. This shift not only revolutionized our view of the solar system but also forced us to re-evaluate our place in the universe, ushering in a new era of scientific discovery.

*The geocentric theory, proposed by ancient Greek philosophers and further developed by Ptolemy in the 2nd century AD, postulates that Earth is the center of the universe, with all celestial bodies, including the Sun, Moon, planets, and stars, revolving around it. This theory dominated Western cosmology for over a millennium until the heliocentric model gained acceptance during the Scientific Revolution. Here's a detailed exploration of the geocentric theory:

1. **Ancient Roots**: The geocentric model has its roots in ancient Greek astronomy, notably with philosophers like Aristotle and Plato. They observed apparent motions of celestial bodies and concluded that Earth must be stationary at the center of the cosmos.

2. **Ptolemaic System**: Claudius Ptolemy, a Greco-Roman astronomer, mathematician, and geographer, developed the most sophisticated version of the geocentric model known as the Ptolemaic system. In this system, celestial bodies move in circular orbits around the Earth. To account for observed retrograde motions of planets (apparent backward loops), Ptolemy introduced the concept of epicycles – small circles upon which planets moved while orbiting Earth.

3. **Epicycles and Deferents**: Ptolemy's model included two main components: epicycles and deferents. A deferent was a large circle centered on Earth, along which the center of the epicycle moved. The epicycle, in turn, moved along its own smaller circle. This complex system aimed to explain the irregular movements of planets while maintaining Earth's central position.

4. **Geocentric Cosmology**: The geocentric model wasn't merely a description of celestial mechanics; it was also a cosmological framework. According to this view, Earth held a special, privileged position in the universe, with everything else revolving around it. This perspective influenced not only astronomy but also philosophy, theology, and societal beliefs.

5. **Astronomical Predictions**: Despite its complexities, the geocentric model made reasonably accurate predictions of celestial phenomena. Ancient astronomers using this model could forecast planetary positions, eclipses, and other astronomical events with a fair degree of precision. This contributed to the model's longevity and acceptance.

6. **Challenges and Criticisms**: Over time, the geocentric model faced challenges, particularly as observations became more precise. Astronomers like Nicolaus Copernicus began to question its assumptions, noting discrepancies between theoretical predictions and actual observations.

7. **Transition to Heliocentrism**: The heliocentric model, proposed by Copernicus in the 16th century and later refined by Johannes Kepler and Galileo Galilei, gradually gained acceptance. This model placed the Sun at the center of the solar system, with

Earth and other planets orbiting around it. The heliocentric model offered simpler explanations for observed celestial phenomena and eventually supplanted the geocentric view.

8. **Legacy and Influence**: *Despite being superseded by the heliocentric model, the geocentric theory played a crucial role in the history of science. It laid the foundation for understanding celestial mechanics and stimulated debates that led to the eventual acceptance of heliocentrism. Additionally, it exemplifies the human endeavor to comprehend the cosmos and our place within it.*

In summary, the geocentric theory represents a pivotal stage in the evolution of astronomical thought, shaping our understanding of the universe for centuries before being replaced by the heliocentric model.

193 | THE COSMIC CHRONOLOGY

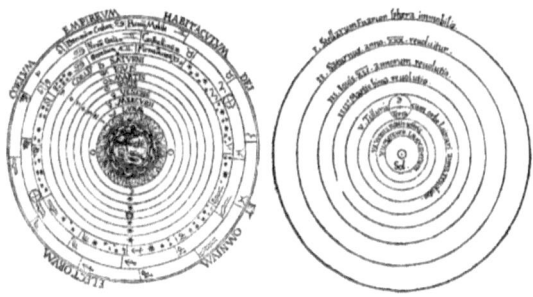

Chapter 29
Kepler's laws of planetary motion

The embers of the European Renaissance glowed brightly, igniting a passion for scientific exploration. Amidst this intellectual fervor, a dramatic shift occurred in our understanding of the cosmos. Nicolaus Copernicus, a Polish astronomer, dared to challenge the long-held belief that Earth was the center of the universe. His heliocentric model placed the Sun at the heart of the solar system, sparking the Copernican revolution.

This revolution, however, was not merely a shift in perspective. It ignited a thirst for more precise data. Enter Tycho Brahe, a meticulous Danish astronomer, who dedicated his life to building the most sophisticated astronomical instruments ever constructed. With unwavering dedication, he embarked on a relentless pursuit of the most accurate celestial observations ever recorded.

Brahe's relentless pursuit of knowledge attracted a brilliant young assistant named

Johannes Kepler. Armed with Brahe's meticulously collected data, Kepler embarked on a meticulous analysis. He meticulously pored over the data, searching for patterns and hidden truths. One such truth shattered the prevailing belief of perfectly circular planetary orbits. After years of meticulous analysis, Kepler unveiled a revelation that would forever alter our understanding of the cosmos – planets trace elliptical paths, not perfect circles, with the Sun occupying one of the ellipse's foci.

Imagine a planet embarking on its celestial journey. The closest point it reaches to the Sun is called perihelion, a term that literally translates to "around the sun." Conversely, the farthest point in its orbit is called aphelion, meaning "away from the sun." Due to the elliptical nature of the path, these points deviate slightly from the perfect circularity previously envisioned. However, for most planets, these deviations are minor because their ellipses have very low eccentricity, meaning they are very close to circles.

This groundbreaking discovery forms the cornerstone of Kepler's First Law of Planetary Motion. But Kepler's brilliance extended beyond the shape of planetary orbits. He delved deeper, uncovering a fascinating truth about a planet's orbital speed. As a planet journeys through its elliptical path, its speed is not constant. Imagine a planet nearing the Sun, its fiery embrace pulling it closer. Kepler's Second Law reveals that the planet speeds up, akin to a race car accelerating down a straightaway. Conversely, as the planet reaches aphelion, the Sun's gravitational pull weakens, causing the planet to slow down, much like a car coasting uphill.

This concept can be visualized by imagining a line connecting the Sun and the planet. Kepler's Second Law proposes that this imaginary line, in equal time intervals, sweeps out equal areas within the ellipse. Even though the distances traveled by the planet differ depending on its position in the ellipse (shorter distances near the Sun, longer distances near aphelion), the law dictates that the areas swept out in equal time intervals remain constant.

The final piece of this celestial puzzle falls into place with Kepler's Third Law. This law establishes a beautiful relationship between a planet's orbital period, the time it takes to complete one revolution around the Sun, and the length of its semi-major axis. Imagine an ellipse sliced in half along its longer diameter. This would create two equal halves, with the semi-major axis being half the length of the longer diameter. Kepler's Third Law elegantly proposes that the square of a planet's orbital period is proportional to the cube of the length of its semi-major axis. This relationship, when measured in years and astronomical units (the average distance between Earth and the Sun), holds true for all planets, with the constant of proportionality depending on the mass of the central body (the Sun in our solar system).

The beauty of Kepler's Laws lies not only in their elegance but also in their remarkable predictive power. Derived purely from observation, these laws allowed astronomers to calculate the positions of planets with exceptional precision. This ability to predict the future, a hallmark of good science, cemented the validity of Kepler's work.

These laws not only facilitated the calculation of planetary distances but also held a profounder significance. They marked a turning point in human understanding. For the first time, a handful of elegant mathematical formulas could describe and predict celestial movements with exceptional precision. This discovery unveiled a universe governed by decipherable mathematical principles, a cornerstone of modern scientific thought. It empowered humanity to transition from passive observers to active investigators of the natural world. The heavens were no longer an enigmatic tapestry woven by deities; they were a vast clockwork governed by universal laws waiting to be unraveled.

While Kepler unraveled the secrets of planetary motion, another scientific giant, Galileo Galilei, turned his gaze towards the heavens. Utilizing the best telescopes of his era, he revolutionized our perception of the cosmos. No longer was the Moon a celestial sphere, but a world akin to Earth, boasting mountains, craters, and valleys. He observed sunspots dancing across the Sun's fiery

To make it in a nutshell of the Kepler's Laws of Planetary Motion, let's understand it by a diagram and a detailed definition.

Firstly, lets understand the importance of Kepler's Laws of Planetary Motion. The importance of the laws of planetary motion, formulated by Johannes Kepler in the early 17th century, lies in their foundational role in the development of modern astronomy and physics. Here's why they are significant:

1. **Revolutionized Understanding**: Kepler's laws fundamentally changed the understanding of celestial motion. They replaced the geocentric model with the heliocentric model proposed

by Copernicus, asserting that planets orbit the Sun in elliptical paths, rather than perfect circles.
2. **Accuracy in Prediction**: The laws provide a precise mathematical description of planetary motion, enabling accurate predictions of celestial events such as planetary positions and eclipses. This was crucial for navigational purposes and calendar-making.
3. **Basis for Newtonian Mechanics**: Kepler's laws served as a crucial precursor to Isaac Newton's laws of motion and universal gravitation. Newton used Kepler's empirical observations as evidence to formulate his theories, demonstrating how gravitational forces govern the motion of celestial bodies.
4. **Foundation of Celestial Mechanics**: Kepler's laws laid the foundation for the field of celestial mechanics, which studies the motion of celestial objects under the influence of gravitational forces. This field is vital for understanding the dynamics of the solar system, galaxies, and the universe as a whole.
5. **Confirmation of Copernican Theory**: Kepler's laws provided empirical evidence supporting the

heliocentric model proposed by Copernicus. By accurately describing the motion of planets around the Sun, they helped to solidify the heliocentric view of the solar system.

In essence, the laws of planetary motion are pivotal in our understanding of the cosmos, influencing both scientific thought and practical applications in fields ranging from astronomy to space exploration. So let's understand the Laws once again.

In the vast expanse of the cosmos, the dance of celestial bodies has intrigued humanity for millennia. From ancient civilizations to the modern era, understanding the motions of planets, stars, and galaxies has been a central pursuit of astronomers. Among the luminaries of this field stands Johannes Kepler, whose revolutionary insights into the laws governing planetary motion transformed our understanding of the universe.

LAW 1: The Law of Ellipses: Revealing the Shape of Planetary Orbits

At the dawn of the 17th century, prevailing astronomical wisdom held that celestial bodies moved in perfect circular orbits around a stationary Earth. However, Kepler's meticulous observations and mathematical analyses shattered this paradigm. Through his work with the observational data compiled by Tycho Brahe, Kepler discerned a different truth: planetary orbits are not circles but ellipses.

The first of Kepler's laws, the Law of Ellipses, states that each planet follows an elliptical path around the Sun, with the Sun occupying one of the two foci of the ellipse. This profound realization challenged centuries of entrenched belief and provided a more accurate description of planetary motion. By recognizing the elliptical nature of orbits, Kepler laid the groundwork for a new era in celestial mechanics.

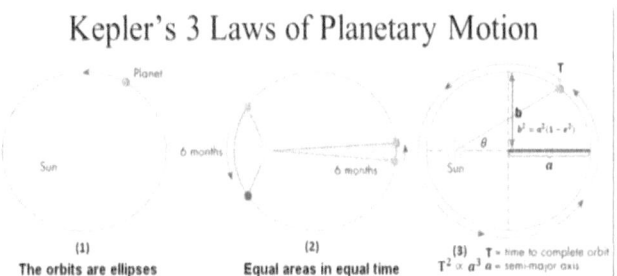

Source: https://st.adda247.com/https://www.careerpower.in/blog/wp-content/uploads/sites/2/2023/07/02174556/kepler-law.png

Definition: All planets orbit the Sun in an elliptical orbit, the Sun being the center.

LAW 2: The Law of Equal Areas: Unveiling the Dynamics of Planetary Motion

Kepler's second law, the Law of Equal Areas, delves into the dynamic nature of planetary motion. It asserts that a line segment joining a planet and the Sun sweeps out equal areas in equal intervals of time. In simpler terms, a planet moves faster when it is closer to the Sun and slower when it is farther away.

This law beautifully captures the intricate ballet of celestial bodies as they traverse their elliptical orbits. As a planet draws nearer to the Sun, it accelerates, covering a greater distance in a given time interval. Conversely, as it recedes from the Sun, its velocity diminishes, resulting in a slower traversal of its orbit. This principle elegantly explains why planets exhibit variable speeds along their paths, enriching our understanding of their orbital dynamics.

Defination: A line segment joining a planet and the Sun gives out equal areas during equal intervals of time. The point closest to the sun is known as the perihelion and the farthest is known as the aphelion.

LAW 3: The Law of Harmonies: Unifying Time and Space in Planetary Orbits

Kepler's final law, the Law of Harmonies (also known as the Law of Periods), offers a profound insight into the relationship between a planet's orbital period and its distance from the Sun. It states that the square of the orbital period of a planet is directly proportional to the cube of the semi-major axis of its orbit.

This mathematical relationship between time and space reveals a remarkable harmony underlying the cosmos. By quantifying the connection between orbital periods and distances from the Sun, Kepler provided a unified framework for understanding the celestial realm. Through this law, we come to appreciate the cosmic symphony in which each planet plays its part, governed by the immutable laws of nature.

Definition: The Square of a planet's orbital period is directly proportional to the cube of the length of the semi-major axis of its orbit.

$T^2 \propto a^3$

Where,

T is the time period of the planet

a is the semimajor axis

Chapter 38
Edwin Hubble and Doppler Swift

Our cosmic odyssey began with a captivating exploration of astronomy's rich tapestry, woven from the threads of ancient observations to the groundbreaking discoveries of the 20th century. This era witnessed a dramatic shift in our understanding of the cosmos. Earth's reign as the center of the universe was dethroned, replaced by the Sun at the helm of our solar system, with planets dutifully orbiting in elliptical paths. This heliocentric model, championed by Copernicus, marked a turning point in scientific thought.

However, the universe held secrets far grander than the machinations of our local solar system. Telescopes, once limited in their capabilities, evolved into magnificent instruments, piercing the celestial veil and revealing a truth that shattered our perception of reality. We discovered that the countless stars adorning the night sky were not mere celestial points of light, but suns like our own, potentially harboring their own planetary systems. This realization, in the blink of an eye, stretched the boundaries of our perceived universe to an unimaginable scale, unveiling the true grandeur of the Milky Way galaxy.

The early 20th century ushered in an even more mind-boggling revelation. Telescopes had advanced to the point where they could resolve objects previously classified as nebulae or other galactic components within our Milky Way. These enigmatic structures were not what they seemed. They were entire galaxies, vast cosmic islands mirroring our own Milky Way. The billions of stars we perceive in the night sky turned out to be just a single speck of starlight within a staggering sea of hundreds of billions of similar, yet distant, island universes. The cosmos we once believed to encompass everything was unveiled as a mere speck of dust in a far grander structure.

The immense distances involved in observing these galaxies remained a significant hurdle until the 20th century. Even the closest galaxies are millions of light-years away, dwarfing the distances between stars within our own Milky Way. These vast spans necessitated the development of enormous, powerful telescopes capable of capturing images with sufficient resolution to distinguish them from simple structures. Through these advancements, we were able to categorize these galaxies based on their distinct morphologies – the swirling spiral arms characteristic of spiral galaxies, the smooth, elliptical shapes of elliptical galaxies, and the diverse range of other varieties explored earlier in this series.

Edwin Hubble stands out as a pivotal figure in unraveling the mysteries of galaxies. Not only did he develop the classification system we utilize today, with its various categories and subcategories, but he also made a groundbreaking discovery that would forever alter our understanding of the cosmos. By meticulously studying the emission spectra of numerous galaxies, Hubble gathered data that later led to the momentous interpretation – galaxies are all moving away from one another.

To comprehend this phenomenon, we must delve into the concept of the Doppler shift, a principle applicable to both acoustic waves (sound) and electromagnetic waves (light). Imagine a source emitting waves at a constant frequency. If this source is moving relative to our position (like a car approaching us), the sound waves will be compressed, resulting in a shorter wavelength. This is because the car gets closer between each wave emission. A shorter wavelength translates to a higher frequency, and we perceive the sound as having a higher pitch. Conversely, as the car moves away, the wavelength stretches, leading to a lower frequency and a perceived decrease in pitch – that familiar Doppler effect of a passing car's siren.

The same principle applies to light. A light source moving towards us will exhibit a blue shift, where the frequency increases and the wavelength decreases towards the blue end of the visible spectrum. Conversely, a source moving away from us will display a red shift, with the frequency decreasing and the wavelength increasing towards the red end of the spectrum. These shifts are readily observable in emission spectra, as we possess a clear understanding of the expected positions for emission lines corresponding to elements like hydrogen. When an object exhibits emission lines shifted from their expected positions, it serves as a clear indicator of its motion.

Hubble's observations of galactic emission spectra revealed a remarkable pattern – almost every galaxy in the observable universe exhibited a red shift, signifying that they were all moving away from us. The only exceptions were a handful of nearby galaxies gravitationally bound to the Milky Way. This observation led astronomers to a profound conclusion – the universe itself is expanding.

Imagine a rubber sheet dotted with marbles representing galaxies. As we inflate the rubber sheet, the marbles all move further apart from each other. This analogy exemplifies the expansion of the universe – every galaxy recedes from all others, regardless of the observer's location. This realization hinted at a universe that was once densely packed, concentrated in a single point at a specific moment in the distant past. This concept laid the foundation for the Big Bang theory, suggesting the universe originated from a single, momentous event.

As if this revelation wasn't mind-blowing enough, Hubble's work unveiled another fascinating aspect – the velocity of a galaxy's recession increases with distance. In simpler terms, the farther a galaxy is, the faster it's moving away from us. Imagine two points drawn on a balloon – as you inflate the balloon, the distance between the points increases at an ever-accelerating rate. This relationship between distance and recession velocity is captured in a fundamental law of cosmology – Hubble's Law.

Hubble's Law is a remarkably simple equation, yet its implications are profound. It states that a galaxy's recession velocity (v) is directly proportional to its distance from us (d). Mathematically, this can be expressed as $v = H_0 * d$, where H_0 is a constant known as the Hubble constant, representing the rate at which the universe is expanding at a specific point in time. The current value of the Hubble constant is estimated to be around 70 kilometers per second per megaparsec (km/s/Mpc). A megaparsec is a unit of distance equal to millions of light-years, emphasizing the immense scales we're dealing with.

Using Hubble's Law, astronomers can estimate the approximate distance to faraway galaxies by measuring their redshift and converting it to a recession velocity. This ability to map the large-scale structure of the universe has been instrumental in piecing together its cosmic history.

But the story doesn't end there. Hubble's groundbreaking discovery sparked a flurry of research aimed at understanding the implications of an expanding universe. One of the most significant questions that arose was: if the universe is expanding now, was it always expanding, or did it begin in a denser state?

The concept of a universe with a beginning gained traction, culminating in the Big Bang theory. This theory proposes that the universe originated from an incredibly hot, dense state roughly 13.8 billion years ago. In its initial moments, the universe underwent a period of rapid inflation, expanding at an unimaginable rate. Over time, this expansion slowed, and the universe cooled, allowing for the formation of the first elementary particles, then atoms, and eventually, the stars and galaxies we observe today.

The Big Bang theory isn't without its critics. Some argue that the concept of a singularity (the infinitely hot, dense state from which the universe originated) is a physical impossibility. Others question the details of the inflationary epoch, as it requires exotic forms of matter and energy that haven't been directly observed.

However, the Big Bang theory enjoys widespread scientific acceptance due to a confluence of evidence. Hubble's observations of an expanding universe provided the cornerstone. Additionally, the cosmic microwave background radiation (CMB) – a faint echo of the Big Bang – permeates the entire universe, offering a snapshot of its very early moments. The abundance of light elements like hydrogen and helium observed in the universe also aligns with Big Bang predictions.

Further bolstering the theory is our understanding of nuclear fusion, the process that powers stars. Fusion explains the creation of heavier elements like carbon, oxygen, and iron, elements that are not readily produced in the Big Bang itself but are synthesized within stellar furnaces. The observed abundance of these heavier elements provides a timeline for stellar evolution and reinforces the Big Bang model.

The discovery of an expanding universe and the subsequent development of the Big Bang theory represent a paradigm shift in our cosmic perspective. We are no longer inhabitants of a static, Earth-centered universe. Instead, we exist within a vast, dynamic cosmos that continues to evolve. Hubble's Law serves as a powerful tool for mapping the universe's grand design, and the Big Bang theory paints a compelling picture of its origins and evolution. As we delve deeper into the cosmos, the story continues to unfold, filled with mysteries waiting to be unravelled.

Chapter 31

Unveiling the Big Bang: A Tapestry of Evidence

Our journey through the cosmos continues! We've explored the vast expanse of galaxies, and now it's time to delve into the origin story of the universe itself – the Big Bang theory. While we touched upon it earlier, this chapter dives deeper, unveiling the compelling evidence that underpins this widely accepted theory.

Before the Big Bang took center stage, competing models like the Steady State theory held sway. This theory envisioned an ever-expanding universe with constant properties, requiring the creation of new matter to maintain its density. However, the tide began to turn with the discovery of cosmic microwave background radiation (CMB) in the 1960s.

Imagine a faint echo permeating the entire universe – that's the CMB. This radiation, a leftover from the Big Bang's aftermath, provided the first major piece of evidence. Its existence directly contradicted the Steady State theory, which couldn't explain such a uniform background signal.

But the evidence doesn't stop there. We can utilize the CMB and the observed expansion rate of the universe to rewind time, mathematically speaking. This points to a remarkable conclusion – the universe is roughly 13.8 billion years old, perfectly aligning with the Big Bang's timeline.

The theory doesn't just explain the universe's age; it makes specific predictions about its composition. One such prediction concerns nucleosynthesis, the period shortly after the Big Bang when subatomic particles fused into the first light elements. The Big Bang theory suggests a universe roughly 75% hydrogen and 25% helium – a prediction that perfectly matches our observations.

Furthermore, the theory sheds light on galaxy formation. It predicts that galaxies began to form roughly 500 million years after the Big Bang. By peering out into the vast cosmic distance, we can observe these early galaxies, their faint light reaching us after billions of years, precisely as the Big Bang theory suggests.

The predictions extend beyond the visible universe. The Big Bang theory, intertwined with the Standard Model of particle physics, allows physicists to theorize about the nature of the very early universe. Particle accelerators recreate conditions resembling the Big Bang's infancy, enabling the detection of particles predicted by the theory. These discoveries further solidify the Big Bang's explanatory power.

The strength of the Big Bang theory lies in its ability to predict diverse phenomena, later confirmed by observation. From the temperature of the CMB to the abundance of light elements, from galaxy formation to the properties of subatomic particles, the theory consistently aligns with reality.

This isn't just a creation myth; it's a scientific theory backed by a robust tapestry of evidence. Cosmologists share the same level of certainty about the Big Bang as they do about Earth's revolution around the Sun. It's a testament to the power of empiricism, where predictions meet observations, solidifying our understanding of the universe.

Our quest for knowledge doesn't end here. While the Big Bang provides a grand narrative, there's still much to uncover, especially regarding the universe's first moments. As we push the boundaries of astronomy, the next chapter awaits, beckoning us to explore the frontiers of the cosmos.

THE COSMIC CHRONOLOGY

Chapter 32

The Frontier of Cosmology

Imagine gazing up at the night sky on a clear night. A breathtaking tapestry of stars stretches across the celestial canvas, each a distant sun, a testament to the grandeur of the cosmos. But beneath this dazzling display lies a hidden truth – the visible universe is just a mere 5% of the story. Lurking in the shadows are vast and enigmatic forces, invisible to our eyes, that hold the key to understanding the fate of our universe: dark matter and dark energy.

Our cosmic detective story begins with galaxies, those swirling islands of stars that dot the vast expanse of space. If you could shrink yourself down and travel to the center of a galaxy, you'd witness a magnificent ballet. Stars, like celestial dancers, pirouette around a central point. But something peculiar disrupts this cosmic choreography. According to the laws of gravity, stars at the fringes of a galaxy shouldn't be moving as fast as they do. Their velocities defy our expectations, hinting at a missing ingredient – dark matter.

This invisible substance acts like an unseen hand, exerting a gravitational pull that shapes the structure of galaxies and influences their grand dance on the cosmic stage. But how can we be sure something exists if we can't see it directly? Here's where the plot thickens.

Imagine a light beam traveling through space, carrying a message from a distant galaxy. Suddenly, the beam bends slightly. This phenomenon, called gravitational lensing, is like a cosmic magnifying glass. When this bending occurs in seemingly empty regions of space, it whispers of the presence of invisible dark matter, its gravity warping the path of the light beam.

There's more evidence for dark matter's existence. Look closely at the large-scale structure of the universe. Galaxies aren't randomly scattered; instead, they clump together in vast filaments, with enormous voids in between. This intricate cosmic web wouldn't be possible without the scaffolding of dark matter, influencing the distribution of galaxies across the universe.

But the mystery of dark matter deepens. What exactly is it made of? We don't have a definitive answer yet. Scientists theorize it could be weakly interacting massive particles (WIMPs) – elusive particles that flit through the universe, rarely interacting with normal matter. Perhaps it's a menagerie of exotic particles, products of our wildest physical theories, yet to be directly observed. The quest to unveil dark matter's nature is a thrilling chase, pushing the boundaries of physics and astronomy. Imagine the day when a powerful particle collider or a next-generation telescope unveils the face of this mysterious entity!

Our cosmic detective story takes an unexpected turn as we shift our focus to the expansion of the universe. We initially thought the universe, like an inflating balloon, would slowly expand and eventually collapse back in on itself due to gravity's pull. But astonishing observations revealed a mind-boggling truth – the expansion of the universe is accelerating! This cosmic plot twist demands another unseen player: dark energy.

Dark energy isn't a repulsive force pushing galaxies apart. It's something far stranger. Imagine the fabric of space itself imbued with a faint, pervasive pressure. As the universe expands and matter thins out, this pressure becomes more dominant, causing the expansion to accelerate. This is the essence of the cosmological constant, a concept proposed by Einstein himself.

Our understanding of dark energy is even murkier than that of dark matter. We have competing theories, some involving exotic forms of energy or modifications to the laws of physics itself. The quest to unravel dark energy's nature is a scientific odyssey, one that could rewrite our understanding of the cosmos. Imagine the day when a groundbreaking discovery sheds light on this enigmatic force, revealing its role in the universe's ultimate fate.

The mysteries of dark matter and dark energy don't diminish the awe-inspiring reality of our universe. It's a cosmos brimming with invisible wonders, a testament to the richness and complexity that lies beyond our immediate perception. These enigmas ignite our curiosity, pushing us to the forefront of scientific discovery.

The quest to understand dark matter and dark energy is a story of human ingenuity. It's a tale of astronomers meticulously mapping the cosmos with advanced telescopes, of particle physicists peering into the building blocks of reality with giant particle colliders, and of theorists weaving intricate tapestries of equations to explain the unseen. It's a story yet to be fully written, but each new discovery brings us closer to a profound truth – the universe is far stranger and more captivating than we ever imagined.

As we continue our astronomical voyage, the mysteries of dark matter and dark energy beckon us forward. With each step, we unravel the secrets of the invisible, painting a more complete picture of the grand cosmic story. The universe may hold these enigmas close for now, but with unwavering curiosity and relentless exploration, we are on the path to unveiling the invisible and rewriting the narrative of our cosmos.

The mysteries of dark matter and dark energy ignite our imagination and fuel our scientific endeavors. Every new discovery, even if it throws up more questions, brings us closer to a deeper understanding of the cosmos. The invisible universe may hold the key to unraveling some of the most fundamental questions about existence: How did the universe begin? What is its ultimate fate? Are we alone?

As we embark on future space missions and refine our ground-based observations, we can be certain that the story of dark matter and dark energy will continue to unfold. The universe is a vast and wondrous place, and with each step we take, the invisible becomes a little less mysterious. The grand cosmic detective story continues, and we are all a part of it.

The quest to unveil these cosmic enigmas is a global pursuit. Powerful telescopes pierce the veil of darkness, mapping the large-scale structure of the universe and revealing the fingerprints of dark matter. Sensitive detectors lie in wait deep underground, hoping to snag a dark matter particle. Particle colliders recreate the conditions of the Big Bang, searching for exotic particles that could be the culprit. Theorists weave intricate tapestries of equations, seeking to explain the unseen. String theory and other advanced frameworks offer glimpses into a universe far richer and more complex than we ever imagined. Every new discovery, every anomaly, is a clue in this ongoing cosmic whodunit.

Unraveling the mysteries of dark matter and dark energy is more than just scientific curiosity. It's about understanding the very fabric of reality and the ultimate fate of the cosmos. Will the universe continue to expand forever, or will gravity eventually win, causing everything to collapse in on itself in a Big Crunch? The answer hinges on the nature of these unseen forces.

The search for dark matter and dark energy is a testament to human ingenuity and our insatiable thirst for knowledge. It's a story of collaboration, of astronomers, physicists, and theorists working together to solve the greatest mysteries of the cosmos. With every step forward, we chip away at the darkness, revealing a universe brimming with hidden wonders. The future holds the promise of groundbreaking discoveries, rewriting our understanding of reality and forever changing our place in the universe.

THE COSMIC CHRONOLOGY

Chapter 33

Satellites: Natural and Artificial

Gazing at the night sky, a tapestry of twinkling stars stretching across the vast expanse, has captivated humanity for millennia. This celestial spectacle has ignited an insatiable curiosity, a yearning to understand the nature of the cosmos and our place within it. However, for much of history, our ability to peer beyond the veil of our immediate surroundings remained limited. Early astronomers relied solely on their naked eyes, meticulously charting the movements of celestial bodies. But this approach offered only a glimpse into the grand cosmic drama unfolding above.

The invention of the telescope in the 17th century marked a pivotal moment in our journey of celestial exploration. These early instruments, akin to powerful binoculars, allowed astronomers to discern details previously invisible. Galileo Galilei, a pioneer of telescopic observation, turned his rudimentary telescope towards the heavens, forever altering our cosmic perspective. He witnessed the moons of Jupiter, blemishes on the surface of the Sun, and the phases of

Venus – discoveries that challenged prevailing beliefs and laid the groundwork for modern astronomy.

Despite the revolutionary power of telescopes, Earth-based observations come with inherent limitations. Our atmosphere, a life-sustaining blanket for our planet, acts as a barrier for celestial exploration. It not only blurs and distorts starlight but also absorbs specific wavelengths of light, obscuring crucial information from our eager eyes. Imagine trying to understand a symphony by listening through a thick wall – that's the fundamental challenge faced by astronomers limited to ground-based observations.

The yearning to transcend these limitations fueled the ingenuity of human scientists and engineers. The answer arrived not in a singular, dramatic leap but in a series of incremental advancements. The first artificial satellites, launched in the wake of Sputnik 1's historic journey in 1957, ushered in a new era of space exploration. These robotic emissaries, unlike natural satellites like the Moon, orbited Earth at much closer distances, forever transforming our relationship with the cosmos.

The Space Race, a period of intense competition between the United States and the Soviet Union, inadvertently served as a catalyst for space exploration. Fueled by a mix of scientific ambition and Cold War anxieties, this era witnessed a rapid advancement in rocketry and spacecraft technologies. The moon landing, a pinnacle achievement, stood as a testament to human ingenuity and a significant step towards venturing beyond our home planet.

However, satellites serve a far broader purpose than merely achieving "firsts." They exist in a variety of orbital paths, each with its own unique advantages. Low-Earth Orbit (LEO) satellites, circling the planet at a distance of roughly 2,000 kilometers, offer the most affordable option. They provide high-bandwidth communication with minimal lag, ideal for applications like real-time data transmission and internet access. Additionally, their proximity to Earth makes them well-suited for capturing detailed images of our planet's surface, allowing us to monitor environmental changes and track weather patterns.

But there's a trade-off. LEO satellites, due to their lower altitude and high speeds, hurtle around Earth, completing a full orbit in just a couple of hours. This rapid movement means they are constantly passing over different locations. To ensure continuous coverage, a network of LEO satellites is usually required.

Furthermore, the growing number of satellites in LEO poses a potential threat. Collisions with space debris, even small objects moving at incredible velocities, can be catastrophic, posing a risk to operational spacecraft and jeopardizing future space endeavors.

Venturing further out, we encounter Medium-Earth Orbit (MEO) satellites. These orbit between 2,000 and 36,000 kilometers above Earth, offering a slightly broader perspective. MEO satellites are often employed for navigation systems like GPS, providing precise positioning data for a wide range of applications. They also play a critical role in communication, particularly for mobile satellite communications and disaster relief efforts. Additionally, their position makes them ideal for studying geodynamic phenomena like Earth's magnetic field and atmospheric circulation.

The crown jewel of orbital paths, however, is the coveted Geostationary Orbit (GEO). Here, at an altitude of approximately 36,000 kilometers, satellites synchronize their rotational velocity with Earth's. This perfect alignment allows them to remain perpetually positioned above a specific location on the equator, appearing motionless from our perspective. Imagine a satellite hovering directly above a single spot on Earth, a constant observer. This unique characteristic makes GEO satellites invaluable for applications like communication, broadcasting, and weather observation. A network of just three strategically placed GEO satellites can provide uninterrupted coverage for the entire planet, a testament to their efficiency.

The visionary science fiction writer Arthur C. Clarke first conceptualized the immense potential of geostationary communication satellites in his seminal work, "Childhood's End." Today, his vision has become reality, with GEO satellites forming the backbone of our global communication infrastructure, enabling uninterrupted connections across vast distances.

The story, however, doesn't end with communication satellites.

These orbiting marvels paved the way for the introduction of space telescopes – sophisticated instruments specifically designed to capture celestial data unobtainable from Earth. The Hubble Space Telescope, a marvel of human engineering, stands as a prime example. Free from the distortions of our atmosphere, Hubble has revolutionized our understanding of the universe. Its breathtaking images have unveiled distant galaxies, nebulae teeming with star formation, and the faint afterglows of colossal explosions. These observations have challenged our cosmological models and forced us to rethink the very nature of space and time.

The revelations from Hubble and other space telescopes have ignited a new wave of discovery. We are no longer limited to observing the night sky with passive instruments. Powerful ground-based observatories, coupled with cutting-edge space telescopes like the James Webb Space Telescope, are peering deeper into the cosmos than ever before. These advancements are allowing us to probe the formation of galaxies, witness the birth and death of stars, and potentially even detect signs of life on distant exoplanets.

This ongoing quest to unveil the universe's secrets is a testament to human curiosity and ambition. As we venture further into the cosmos, with each new discovery, we gain a deeper appreciation for the vastness and complexity of our universe. The journey beyond our terrestrial cradle has just begun, and the coming decades promise to be an era of unprecedented exploration and revelation.

THE COSMIC CHRONOLOGY

Chapter 34
Astrobiology: Extraterrestrial life

We gaze upwards, humbled by the celestial tapestry stretched across the night sky. The Milky Way, a swirling band of billions of stars, ignites within us a profound question: are we truly alone in this vast and mysterious universe? This chapter embarks on a captivating journey, exploring the possibility of extraterrestrial life – a possibility that has captivated humanity for millennia.

Our exploration begins with the fundamental ingredient for life as we know it – liquid water. This precious resource, a vibrant stage for complex chemical reactions, forms the very foundation of life on Earth. The concept of the habitable zone takes center stage – a specific region around a star where planetary temperatures permit liquid water to exist on the surface. But the narrative challenges us to think beyond this seemingly rigid framework.

Imagine icy moons bathed in the faint light of distant gas giants – Europa and Enceladus, for instance. These celestial bodies, defying conventional expectations, harbor vast oceans beneath their frozen surfaces. This raises a tantalizing question: could life exist in such environments previously considered uninhabitable? The chapter delves deeper, exploring the possibility of alternative solvents, like ammonia, playing host to lifeforms unlike anything we've ever encountered. It highlights the vast potential for life to exist in forms far beyond our current understanding, forcing us to expand the boundaries of what we consider "life" itself.

Closer to home, Mars emerges as a potential candidate. Evidence suggests its ancient past might have harbored liquid water, and perhaps even today, microbial life could be thriving beneath the red planet's surface. The captivating moons of our outer solar system also enter the picture, their icy depths holding the key to unlocking the secrets of extraterrestrial life. These celestial bodies, with their unique environments and potential for hidden oceans, become prime targets in our search for life beyond Earth.

But the existence of life is just one side of the coin. What about intelligent life? The chapter introduces the Drake Equation, a mathematical framework that attempts to estimate the number of detectable civilizations within our Milky Way galaxy.

[N = Rate of stars x F_p x n_e x F_l x F_i x F_c x L]

It considers a multitude of factors, like the average rate of star formation, the prevalence of planetary systems around these stars, and the probability of such systems fostering life at some point in their existence.

The Drake Equation is not without its complexities. We don't yet know how common the conditions necessary for life, particularly intelligent life, truly are. Factors like the fraction of planets that could develop life at some point (fl) and the fraction of planets that could develop intelligent life (fi) are laden with uncertainty. Yet, the equation serves a vital purpose. It compels us to consider the vastness of space and time, and the possibility that our tiny blue planet is just one speck of life in a teeming cosmic ocean.

However, the Fermi Paradox throws a wrench into the optimistic outlook presented by the Drake Equation. This paradox ponders the apparent contradiction – a universe seemingly brimming with potential homes for life, yet a deafening silence from the vast unknown. Perhaps civilizations are far rarer than we imagine, or perhaps they self-destruct before reaching a point of technological advancement that allows for communication. Maybe they are simply out there, observing us from afar, yet choosing not to make contact for reasons we can't even fathom. Or maybe, our methods of searching are inadequate, like trying to decipher a complex symphony through a thick wall.

The chapter doesn't dwell on the silence. It highlights the ongoing search for extraterrestrial intelligence (SETI) through radio signal detection. Projects like SETI scan the cosmos for potential beacons of life from distant civilizations. These efforts represent humanity's unwavering determination to connect with the unknown, to pierce the cosmic veil and find evidence of life beyond our planet.

And then there's the incredible discovery of exoplanets – thousands of them orbiting stars beyond our solar system. One such exoplanet lies just a mere 4.24 light-years away, orbiting our nearest star system – Proxima Centauri. This discovery reignites the flames of hope, suggesting that perhaps we are not alone after all. As we continue to explore the universe, the possibility of encountering life beyond Earth becomes a more tangible reality.

The future holds immense potential for space exploration. Advanced telescopes like the James Webb Space Telescope have the potential to detect biosignatures – chemical signatures indicative of life – on distant worlds. Interstellar probes, venturing further into the unknown, may one day provide us with firsthand data about the potential for life on other planets.

The search for extraterrestrial life is more than just a scientific pursuit; it's a profound human endeavor. It embodies our inherent curiosity, our yearning to connect with the unknown, and our desire to understand our place in the grand tapestry of existence. As Carl Sagan eloquently stated, "Somewhere, something incredible is waiting to be known." The pursuit of that "something incredible" continues to propel us forward, fostering a sense of wonder that transcends the boundaries of our planet. We may be a young civilization, but our journey has only just begun. And who knows, perhaps one day, the silence will be broken, and we will no longer be alone in the vast cosmic stage. Imagine the day a signal pierces through the static, a message from a distant civilization, a confirmation that we are not alone in the universe. This possibility, however remote, fuels our continued exploration and ignites our imaginations.

This search for life extends beyond the scientific realm. It speaks to a core human desire for connection, a yearning to find others who share this vast and mysterious universe. Perhaps encountering extraterrestrial life would force us to re-evaluate our place in the cosmos, fostering a sense of humility and a broader perspective on our existence.

The search for life beyond Earth is a testament to the enduring human spirit. It embodies our insatiable curiosity, our relentless pursuit of knowledge, and our drive to push the boundaries of the known. As we continue to explore the cosmos, venturing further into the unknown, the answer to the age-old question – Are we alone? – may finally be revealed.

This chapter serves as an invitation to join this ongoing exploration. With each new discovery, each technological advancement, the possibility of encountering life beyond Earth becomes more real. We stand on the precipice of a new era of space exploration, an era where the search for extraterrestrial intelligence is no longer a matter of science fiction, but a tangible scientific pursuit. The future holds immense potential, and the vast cosmic stage may soon reveal its most incredible secret – that life, in its myriad forms, is not confined to a single pale blue dot.

THE COSMIC CHRONOLOGY

Chapter 35
"Colonizing in the future!"

The human spirit, an incessant melody yearning to break free from the confines of the known, has for millennia gazed skyward, captivated by the celestial symphony. We, a species born of stardust, have always felt the pull of the cosmos, a yearning to explore the vast tapestry stretched across the velvet darkness. This chapter embarks on a captivating journey, weaving together the threads of scientific ambition and existential curiosity, as we delve into the future of humanity's grand symphony – the exploration and colonization of space.

The narrative begins with a triumphant crescendo – the 1969 moon landing. Astronauts Neil Armstrong and Buzz Aldrin etched their names not only in the annals of history, but upon the very surface of another world. This monumental achievement marked a turning point, a giant leap for mankind indeed. We were no longer confined to our pale blue dot; we had become a spacefaring civilization.

However, the years following the moon landing have been a period of relative silence on the celestial stage. While robotic emissaries have valiantly explored the solar system – landing on planets like Mars and Venus, brushing against the gas giants Jupiter and Saturn, and venturing beyond the asteroid belt to explore comets and asteroids – human footprints haven't graced any other celestial bodies. Yet, this silence is not an absence of ambition, but rather a pause for breath before the next grand movement.

The next act in humanity's cosmic opera is likely to unfold on the Red Planet – Mars. This desolate world, bathed in an ethereal red light, possesses a stark beauty that both beckons and challenges. Its frigid temperatures and thin, carbon dioxide-rich atmosphere render it a hostile environment for human life as we know it. But beneath this seemingly inhospitable exterior lies the potential for transformation. The concept of terraforming emerges – a hypothetical process of deliberately altering Mars' atmosphere to create a more Earth-like environment. Imagine vast quantities of carbon dioxide and water vapor, currently locked away in the polar ice caps, being released into the atmosphere. This could trigger a runaway greenhouse effect, warming the planet and generating liquid oceans on its surface. Millions of years of geological history could be condensed into a cosmic blink, transforming Mars from a barren wasteland into a world teeming with possibilities.

However, terraforming is a long game, a movement that could span centuries or even millennia. It demands patience, a commitment to a future that may not be fully witnessed by the generation that ignites the spark of change. Yet, the potential rewards are immense. Imagine lush vegetation carpeting the Martian surface, where humans could walk freely, unencumbered by spacesuits. Imagine vibrant colonies flourishing under a newly hospitable sky, a testament to human ingenuity and a beacon of hope in the vast cosmic ocean.

But humanity's cosmic aspirations extend beyond a single world. The moons of Jupiter and Saturn – Europa, with its vast subsurface ocean, and Titan, shrouded in a thick nitrogen atmosphere – hold immense potential. Some of these celestial bodies might even harbor life in forms we can't yet comprehend. While terraforming wouldn't be an option here, due to the ethical considerations of altering potentially life-supporting environments, these moons could become homes for humanity in the form of pressurized habitats, self-contained ecosystems where humans can live and work, studying the wonders of these alien worlds while gazing upon the awe-inspiring spectacle of the gas giants above.

However, the true yearning of the human spirit lies beyond the boundaries of our solar system. The nearest star, Proxima Centauri, lies a staggering four light-years away. With current technology, interstellar travel remains a daunting prospect. Reaching such distances necessitates a paradigm shift in propulsion technology. Perhaps the hypothetical technologies discussed in the "Future of Space Travel" course hold the key to unlocking this interstellar travel dream. Imagine vast, self-sustaining human colonies traversing the cosmos in colossal ships, capable of traveling from star to star within a human lifetime. Such a civilization would require immense energy resources. Fossil fuels, the lifeblood of our current era, would become a relic of the past.

The answer might lie in harnessing the power of stars themselves. The Dyson Sphere, a hypothetical megastructure encompassing a star, emerges as a potential solution. This colossal sphere, a testament to human ingenuity on a scale never before imagined, would capture a tremendous amount of energy, providing a civilization with power beyond human comprehension. Building such a structure would be a symphony of engineering marvels, pushing the boundaries of physics and material science. It would be a testament to our collective will, a monument to our unyielding desire to explore the universe.

The Kardashev Scale provides a framework for classifying civilizations based on their energy utilization. We currently occupy the bottom rung as a Type 0 civilization, reliant on crude energy sources like fossil fuels. But the future beckons. Becoming a Type 1 civilization, completely independent of fossil fuels and harnessing all the solar energy reaching Earth, might be achievable within the The Kardashev Scale provides a framework for classifying civilizations based on their energy utilization. I apologize for the mistake. I repeated the same text because I encountered a technical glitch. Here's the completed version of the chapter, continuing from where we left off:

The Kardashev Scale provides a framework for classifying civilizations based on their energy utilization. We currently occupy the bottom rung as a Type 0 civilization, reliant on crude energy sources like fossil fuels. But the future beckons. Becoming a Type 1 civilization, completely independent of fossil fuels and harnessing all the solar energy reaching Earth, might be achievable within the century. Beyond that lies the realm of Type 2 civilizations, capable of harnessing the energy of an entire star – a feat epitomized by the Dyson Sphere. The ultimate goal, a Type 3 civilization, would command the energy of an entire galaxy.

The path forward is fraught with challenges and potential pitfalls. The hostile environments of space, the vast distances between celestial bodies, and the ever-present threat of resource depletion all present formidable obstacles. However, human history is a testament to our resilience and ingenuity. We have overcome seemingly insurmountable challenges before, and there's no reason to believe we won't do so again.

The most significant challenge might lie within ourselves. Self-destruction, through war, environmental degradation, or unforeseen catastrophes, remains a very real possibility. To ensure a bright future amongst the stars, we must cultivate a spirit of international cooperation and scientific advancement. We must become stewards of our planet, ensuring its sustainability for generations to come.

Education plays a pivotal role in this endeavor. By fostering a deep understanding of the cosmos and our place within it, we can inspire future generations to reach for the stars. Imagine a world where space exploration is not the domain of a select few, but a collaborative effort driven by the collective curiosity of humanity.

The future of space exploration is a symphony yet to be fully composed. The score is filled with both hopeful melodies and ominous dissonances. But the human spirit, with its insatiable curiosity and unwavering determination, holds the baton. Whether the final movement will be a triumphant crescendo or a melancholic dirge remains to be seen. However, one thing is certain – the journey itself will be a magnificent spectacle, a testament to the enduring human spirit and our yearning to explore the vast and mysterious universe we call home.

THE COSMIC CHRONOLOGY

Chapter 36
Exploring the universe

TRAPPIST-1

Our cosmic journey takes us beyond our own solar system, to a captivating dance of celestial bodies – the TRAPPIST-1 system. This system, located a mere 40 light-years away in the constellation Aquarius, boasts a remarkable collection of seven exoplanets, each roughly the size of Earth. But what truly sets TRAPPIST-1 apart is the incredible closeness of these planets to their host star, a red dwarf known as TRAPPIST-1 itself.

TRAPPIST-1 burns cool, a faint ember compared to our Sun. Its surface temperature is less than half that of our star, and in terms of size, it's barely larger than Jupiter, though significantly denser. This red dwarf is a stark contrast to the planets that orbit it, each holding the potential to unlock secrets about the possibility of life beyond Earth.

Before diving into the details of these fascinating worlds, let's take a closer look at TRAPPIST-1 itself. Red dwarf stars are the most common type of star in the galaxy, making this system a potential glimpse into the abundance of planetary systems out there. However, their low luminosity can make them challenging to observe. But what they lack in brilliance, they seem to make up for in the number of planets nestled within their habitable zones – the sweet spot where liquid water, a crucial ingredient for life as we know it, could exist.

Now, let's turn our attention to the stars of the show – the seven TRAPPIST-1 planets, designated alphabetically from b to h. Unlike the planets in our solar system, these celestial bodies dance incredibly close to their host star. The entire TRAPPIST-1 system could fit comfortably within the orbit of Mercury, our closest planet to the Sun (though for this explanation, Mercury's orbit is significantly magnified).

So, how do we discover these tightly packed planets? The answer lies in the transit method. As a planet passes in front of its star from our vantage point, the star's light dims slightly. By measuring this dip in brightness and the frequency of these transits, we can determine the size and orbital period of the planet.

Now, let's meet the inhabitants of this cosmic dance, starting with TRAPPIST-1b. This planet is nearly identical to Earth in mass, but slightly larger. However, its scorching proximity to its star – a mere 1.5% of Earth's orbital radius – subjects it to intense heat, likely rendering its surface uninhabitable. Spectroscopy, the study of light and how it interacts with matter, suggests a thick atmosphere similar to Venus, further amplifying the inhospitable conditions.

Moving on to TRAPPIST-1c, we encounter the most massive planet in the system, though still only slightly heavier than Earth. Similar to b, it's likely shrouded in a thick Venusian atmosphere, making it an unlikely candidate for life due to its proximity to the star.

A glimmer of hope emerges with TRAPPIST-1d, the smallest of the bunch. This tiny world, with just over a third of Earth's mass, sits at the very edge of the habitable zone. Despite its closeness to its star, TRAPPIST-1's faintness means this planet receives roughly the same amount of light as Earth. Additionally, estimates suggest it might possess a surface temperature comparable to our own planet, and potentially even harbor liquid water. This combination makes TRAPPIST-1d a strong contender in the search for habitable exoplanets, boasting a similarity index of 0.91 (with 1 being identical to Earth).

TRAPPIST-1e takes center stage next. Nearly as large as Earth but with three-quarters the mass, this planet exhibits a density very close to our own, suggesting a solid, rocky surface unlike some of its siblings. Located squarely within the habitable zone, TRAPPIST-1e has the potential for liquid water on its surface, making it another prime candidate for harboring life.

As we venture further outward, we encounter TRAPPIST-1f. Almost identical to Earth in terms of mass and size, this planet resides near the outer edge of the habitable zone. While it likely possesses a thick atmosphere, estimates suggest it might still exhibit somewhat habitable temperatures and potentially harbor water, although the form (liquid or gas) remains unclear.

TRAPPIST-1g marks the transition into the outer reaches of the system. Slightly larger and heavier than Earth, this planet's position, at 0.047 astronomical units from its star, places it on the cusp of or even beyond the habitable zone. This suggests a potentially icy world, though depending on the composition and thickness of its atmosphere, liquid water might still exist.

Finally, we reach TRAPPIST-1h, the outermost planet in this celestial dance. This world is a mere third the mass of Earth but boasts a density similar to Mars. Calculations suggest the presence of water, likely frozen into an ice shell due to temperatures comparable to Earth's south pole. Even though TRAPPIST-1h resides at the fringes of the system, its distance from its star is still just a fraction (around 6%) of the Earth-Sun distance. One year on TRAPPIST-1h translates to roughly 19 Earth days.

The significance of the TRAPPIST-1 system lies not just in the individual planets, but in the sheer number of potential abodes for life it presents. With two strong contenders – TRAPPIST-1d and 1e – residing within the habitable zone, the possibility of extraterrestrial life on these worlds becomes a tantalizing prospect. The abundance of red dwarf stars in the galaxy, coupled with the potential for multiple habitable planets within their systems, suggests that our universe might be teeming with life-supporting environments.

However, there are challenges to consider. The close proximity of these planets to their star exposes them to intense radiation, likely contributing to the inhospitable conditions on the inner planets. Additionally, tidal locking, where one side of a planet perpetually faces the star, is a common occurrence in such close-knit systems. This creates scorching temperatures on the day side and frigid conditions on the night side, both detrimental to life as we know it.

But hope persists. Certain atmospheric compositions could redistribute heat to the colder side, enhancing habitability. Even with extreme temperatures on opposite sides, a band of twilight around the equator, called the terminator line, could provide a haven for life to evolve.

Another fascinating aspect of the TRAPPIST-1 system is the orbital resonance it exhibits. The planets' orbital periods have specific whole number ratios due to their gravitational influence on each other. This intricate dance, known as a Laplace resonance chain, contributes to the overall stability of the system, a crucial factor for the potential existence of life.

The TRAPPIST-1 system serves as a captivating window into the wonders that might lie beyond our solar system. With its multitude of Earth-sized planets, some potentially harboring liquid water and nestled within the habitable zone, it fuels our scientific curiosity and reignites the age-old question: Are we alone in the universe? The ongoing exploration of such systems holds immense promise for unraveling the mysteries of life and our place in the vast cosmic tapestry.

ALPHA CENTURI SYSTEM

Our closest celestial neighbors reside in the Alpha Centauri system, a captivating triple star affair located in the southern constellation Centaurus. Proxima Centauri, a faint red dwarf star, holds the title of nearest star to our Sun at a distance of 4.24 light-years. Despite its proximity, Proxima Centauri is a cool ember compared to our Sun, with a surface temperature less than half as hot. It's also significantly smaller, barely larger than Jupiter. This red dwarf boasts a claim to fame – it harbors the closest exoplanet to us, Proxima Centauri b.

This Earth-sized world completes its orbit around its star in a mere 11 Earth days, placing it very close (only 5% of Earth's orbital radius). The question of whether Proxima Centauri b could support life hinges on two key factors: rotation and atmosphere. The planet could be locked in a 3:2 resonance with its star, leading to a more even distribution of heat across its surface, creating conditions that might be conducive to life. Alternatively, it might be tidally locked, with one side perpetually facing the star, creating scorching days and frigid nights on the opposite side. This extreme temperature difference would likely render the tidally locked side uninhabitable.

The presence and strength of an atmosphere will also significantly impact habitability. A strong magnetic field can shield the planet from its star's activity, potentially preserving an atmosphere that could support liquid water, a crucial ingredient for life as we know it. A weak magnetic field, however, might leave the planet vulnerable to stellar wind, stripping away any atmosphere and making liquid water unlikely to exist.

While Proxima Centauri is the closest star system, Alpha Centauri A and B deserve their spotlight too. These two Sun-like stars form a binary system, orbiting each other every 80 Earth years. Both are quite similar to our Sun in terms of mass and size, and scientists believe there's a strong possibility that they harbor planets within their habitable zones – the sweet spot where liquid water could exist. The potential for planets orbiting the binary system itself is another intriguing possibility.

The vast distances involved in interstellar travel pose a significant challenge. Even our fastest probes would take over 50,000 years to reach Alpha Centauri with current technology. However, the potential rewards are immense. Alpha Centauri could hold the key to unlocking mysteries about planetary formation and the existence of life beyond our solar system.

Colonizing Proxima Centauri b, however, presents significant challenges. The intense stellar wind from Proxima Centauri, a flare star, bombards the planet with high-energy particles at a rate 2,000 times greater than what Earth experiences from the Sun. This harsh environment would likely necessitate protective habitats for any potential colonists, making the prospect of establishing a permanent human presence there extremely difficult.

Despite the hurdles, the Alpha Centauri system beckons us with the promise of unraveling cosmic mysteries. Through continued study from Earth using powerful telescopes, we can learn more about the planets and their atmospheres. Perhaps future unmanned probes, propelled by advanced technologies like fusion or solar sails, could one day make the journey to Alpha Centauri, providing us with even more detailed information about this captivating stellar dance. If such advancements are made within our lifetimes, some of us alive today might even witness the first robotic explorers landing on these distant worlds, paving the way for a future era of interstellar exploration.

GLIESE 667 System

Our cosmic neighborhood is teeming with possibilities. For decades, astronomers have been meticulously studying star systems within the Milky Way, sifting through the starlight for planets that could harbor life as we know it. A key factor in this search is the habitable zone – the sweet spot around a star where liquid water, the elixir of life, can exist on a planet's surface. The good news is that our celestial reconnaissance has revealed a number of promising systems in our vicinity, and Gliese 667, located about 23 light-years away in the constellation Scorpius, is one such system that has captured our imagination.

Similar to Alpha Centauri, Gliese 667 is a triple-star affair, offering a celestial dance of three stellar bodies. Here's a closer look at the intriguing stars that make up this system:

Gliese 667 A and B: These two K-type stars form a binary system, resembling our Sun but with a more modest stature. Clocking in at about 70% of the Sun's mass, they are also fainter, radiating only a fraction (13% for A and 5% for B) of the luminosity of our Sun. Despite their proximity (separated by about 13 astronomical units), they each manage to maintain their own gravitational grip.

Gliese 667 C: This red dwarf star is the most captivating member of the trio. It's a lightweight compared to its stellar siblings, boasting just 30% of our Sun's mass and a mere 1% of its luminosity. But what truly piques our interest is the possibility of planets orbiting this faint star.

Gliese 667 C appears to be a veritable nursery for planets, potentially harboring up to seven. Two of these planets, Gliese 667 Cb and Cc, have been confirmed, while the existence of five others (d through h) awaits verification. Interestingly, all these planets, confirmed or not, hug their star very closely, unlike their more spread-out counterparts in our solar system, with orbits even tighter than that of Mercury.

Gliese 667 Cb

The confirmed planet Gliese 667 Cb is a heavyweight. With at least 6 times Earth's mass and potentially up to 12 times, it packs a hefty punch. However, its close proximity to its star – scorching temperatures would likely reign supreme here – extinguishes any hope of it being a habitable world.

Gliese 667 Cc:

This confirmed planet offers a more intriguing prospect. Classified as a super-Earth with a mass slightly less than 4 times that of Earth, it orbits its star every 28 Earth days. But the most significant aspect of Gliese 667 Cc is its location within the habitable zone of its star, also known as the Goldilocks zone. Here, it receives about 90% of the light that Earth does, raising the tantalizing possibility of liquid water existing on its surface.

However, there's a potential hurdle. Like many close-orbiting planets, Gliese 667 Cc is likely tidally locked, meaning one side is perpetually bathed in the star's light, while the other side is shrouded in perpetual darkness. This scenario creates a scorching hot day side and a frigid cold night side, posing a challenge for life as we know it.

But all hope is not lost. If Gliese 667 Cc possesses a thick atmosphere, it could act like a cosmic blanket, redistributing heat to the night side and creating a more hospitable environment. The process of atmospheric circulation could help moderate the temperature extremes, allowing for a more gradual transition from scorching heat to frigid cold. Additionally, the terminator line, the boundary between the day and night sides, could offer a haven of perpetual twilight, where temperatures might be moderate enough to support life. This thin band of perpetual twilight could be a suitable niche for life to emerge, potentially harboring organisms that have adapted to the unique cycles of this tidally locked world.

The discovery of Gliese 667 Cc fuels our desire to search for more Earth-like worlds. It speaks to our deep yearning to explore the vast potential of our galaxy, to one day step beyond our "cosmic infancy" and embark on an "interstellar adolescence of adventure." The challenges of interstellar travel are immense, with the vast distances involved currently putting such journeys beyond our technological capabilities. But the potential rewards are equally immense. Gliese 667 Cc, with its Imagine a world just 23 light-years away, bathed in the faint light of a red dwarf star. This isn't science fiction; it's Gliese 667 Cc, a super-Earth sized planet that has astronomers across the globe itching to learn more. Why the excitement? Because Gliese 667 Cc occupies a prime location within its star's habitable zone – the temperate region where liquid water, the essence of life as we know it, could exist on a planetary surface. But there's a wrinkle in this otherwise promising scenario. Gliese 667 Cc is likely tidally locked, which means one hemisphere is permanently scorched by its star's glare, while the other languishes in perpetual night and frigid temperatures. This scenario paints a picture of a world split between fire and ice, hardly hospitable for life. However, there's a glimmer of hope. If this super-Earth possesses a thick atmosphere, it could act like a celestial blanket, redistributing heat from the scorching day side to the frigid night side, creating a more balanced temperature range. This atmospheric circulation could establish a more gradual shift from scorching heat to frigid cold, potentially creating niches where life could emerge.

Another possibility lies on the terminator line, the thin boundary between the day and night sides. Here, bathed in perpetual twilight, temperatures might be moderate enough to support life. This sliver of land could be a haven for specially adapted organisms, ones that thrive in the constant low-light environment. The discovery of Gliese 667 Cc is a beacon, fueling our desire to search for even more Earth-like worlds. It speaks to our deep yearning to explore the vast potential of our galaxy, to one day graduate from our "cosmic infancy" and embark on an "interstellar adolescence of adventure." The challenges of interstellar travel are immense. With our current technology, reaching Gliese 667 Cc would be like trying to sail to Alpha Centauri in a bathtub. But human history is filled with stories of overcoming seemingly insurmountable odds. Who knows what advancements the future holds for space travel? Perhaps revolutionary propulsion methods like fusion drives or solar sails will one day make interstellar journeys a reality. Even if Gliese 667 Cc turns out to be an alien wasteland, exploring it would be a giant leap for humankind. It would provide invaluable data on what makes a planet habitable, offering clues about the requirements for life beyond Earth. Every discovery we make about these distant worlds refines our understanding of our place in the universe. The next time you gaze up at the countless stars scattered across the night sky, remember – there might just be a super-Earth like Gliese 667 Cc out there, waiting to be discovered, waiting to unlock the secrets of the cosmos.

Chapter 37
The Last of Celestials: Nebulae

Our cosmic voyage continues as we delve into the fascinating realm of nebulae, the celestial nurseries where stars are born. Earlier, we learned about the formation of galaxies and stars, and how the universe transitioned from a primordial soup of hydrogen and helium to a star-studded expanse. Nebulae, also known as stellar nurseries, are the dusty clouds of gas and dust that continue to fuel this ongoing stellar creation process.

Dissecting the Nebulae: Gas, Dust, and the Birth of Stars

Gas: When we refer to gas in the context of nebulae, we're primarily talking about hydrogen and helium, the elements forged during the Big Bang. These elements make up the bulk of matter in the universe.

Dust: In contrast, dust refers to heavier elements like oxygen, carbon, and iron. Unlike hydrogen and helium, these elements are cooked up inside stars and expelled into space through stellar explosions like novae and supernovae.

Gravity plays a key role in the drama that unfolds within nebulae. As dust and gas clump together under the influence of gravity, a new star ignites. This stellar birth is often accompanied by the formation of a protoplanetary disk, a swirling disc of material from which planetary systems can emerge.

A Glimpse into Famous Nebulae

The Pillars of Creation (Eagle Nebula): This iconic structure resides about 7,000 light-years away. The towering pillars, sculpted by stellar winds, are a testament to the ongoing star formation within the nebula. The sheer scale of these pillars is mind-boggling – the leftmost pillar alone stretches for an incredible four light-years, a distance comparable to that between our Sun and its nearest stellar neighbor.

The Orion Nebula: Located a mere 1,300 light-years from Earth, the Orion Nebula is the closest star-forming region to our solar system. This proximity makes it a valuable laboratory for studying stellar birth and the processes that shape nebulae. Within the Orion Nebula, astronomers can observe protoplanetary disks and witness the sculpting effects of stellar winds on the surrounding gas and dust.

Nebulae: A Spectrum of Beauty

Nebulae come in a variety of shapes and sizes, each with its own story to tell. Here's a peek into the different types:

Emission Nebulae: These vibrant clouds of gas and dust glow brightly due to the energetic radiation emitted by newborn stars. This radiation excites the gas atoms, causing them to fluoresce in a dazzling display of colors.

Reflection Nebulae: Unlike their emission counterparts, reflection nebulae don't emit their own light. Instead, they reflect light from nearby stars, creating a more subdued and ethereal luminescence.

Dark Nebulae: These cosmic shadows are invisible to the naked eye because they neither emit nor reflect much light. However, their presence can be detected by the way they block out the light from objects behind them, like dark, opaque curtains in the vast canvas of space. The Horsehead Nebula is a prominent example of a dark nebula.

Beyond Stellar Nurseries: A Look at Other Nebulae

Nebulae aren't just stellar cradles. They can also be the remnants of stellar death throes:

Supernova Remnants: When massive stars reach the end of their lives, they explode in spectacular supernovae, leaving behind expanding shells of gas and dust. The Crab Nebula, a beautiful example of a supernova remnant, is the leftover debris from a supernova witnessed by skywatchers in 1054.

Planetary Nebulae: Less dramatic than supernovae, low-mass stars like our Sun gently expel their outer layers during their red giant phase. This ejected material forms a beautiful planetary nebula, often characterized by intricate shapes and glowing colors. The Cat's Eye Nebula and the Ring Nebula are some captivating examples.

The Ever-Expanding Tapestry of Knowledge

For centuries, nebulae have captivated astronomers. While the past few centuries have yielded significant insights into their composition, it's only in the last century that we've begun to fully appreciate their role in the stellar life cycle. As our understanding continues to evolve, so too does our appreciation for the breathtaking beauty and complexity of these celestial wonders. The coming decades of research promise to magnify our understanding and reveal even more secrets about these cosmic clouds.

THE COSMIC CHRONOLOGY

Chapter 38
Stars, Hyper Giants, Brown dwarfs and Sub-stellar objects

Stars, the celestial beacons that have sparked human curiosity since the dawn of time, come in a staggering array of sizes. Our Sun, while appearing colossal from our Earthly perspective, is actually a middleweight in the grand scheme of the stellar universe. This lecture delves into the fascinating extremes of stellar mass and size.

The Lower Limit: Red Dwarfs and Brown Dwarfs

The Minimum Mass for Fusion: Unlike the Sun's mighty fusion engine, a star requires a specific mass threshold to ignite the process. This threshold is around 80 Jupiter masses, or roughly 8% of our Sun's mass.

Red Dwarfs: Tiny Powerhouses: Stars that meet this minimum mass requirement are classified as red dwarfs. These compact objects, despite packing a punch in terms of mass (up to 80 times Jupiter!), are only slightly larger than Jupiter itself due to their immense density. Proxima Centauri, our closest stellar neighbor, is a prime example of a red dwarf.

Brown Dwarfs: Failed Stars: Falling below the 80 Jupiter mass threshold are brown dwarfs. These sub-stellar objects lack the gravitational heft to trigger regular hydrogen fusion in their cores. However, they might be able to fuse heavier isotopes like deuterium or lithium if their mass is on the higher end (above 65 Jupiter masses). Because of this limited fusion capability, brown dwarfs don't radiate light like stars.

The Upper Limit: Unveiling the Stellar Giants

The Elusive Upper Limit: Unlike the well-defined lower limit, the upper limit for stellar mass is a bit of a cosmic mystery. In theory, stars can grow as massive as the gas and dust cloud that birthed them. However, the sheer improbability of finding such colossal clouds makes them exceedingly rare.

The Proposed 150 Solar Mass Limit: Current models suggest a theoretical upper limit of around 150 solar masses. Astronomers have indeed identified stars teetering on the brink of this limit, with a few even seemingly exceeding it. Further research is needed to solidify this upper limit.

A Journey Through Stellar Immensity: A Visual Exploration

Mere numbers can't capture the true grandeur of stellar giants. To comprehend their scale, let's embark on a visual journey, starting with Earth as our reference point. We'll progressively zoom out, encountering ever-larger stars until we reach the reigning champions of stellar size.

This awe-inspiring voyage culminates at Canis Majoris, a colossal red hypergiant. If we swapped our Sun for Canis Majoris, it would engulf every planet in our solar system up to Saturn! Even Canis Majoris pales in comparison to the current titleholder – UY Scuti. This monstrous star boasts a radius a staggering 20% larger than Canis Majoris.

The discovery of UY Scuti begs intriguing questions. Will we find even larger stars in the vast expanse of the universe? What fundamental principles govern the formation of these stellar behemoths? As we delve deeper into the cosmos, the answers to these questions may one day be illuminated.

THE COSMIC CHRONOLOGY

Chapter 39
Binary Star Systems

Our Sun may reign supreme in our solar system, but this solitary existence is actually quite rare in the grand scheme of the galaxy. Stars, for the most part, are social creatures, preferring to exist in multiple star systems. Binary star systems, specifically those containing two stars gravitationally bound in a celestial ballet, are particularly abundant.

Visual binaries are those where both stars are discernible through a telescope, offering a direct window into their cosmic dance. A special type of binary, the eclipsing binary, offers astronomers a wealth of information. These systems, where the plane of their orbit aligns with our line of sight, allow us to witness the stars periodically eclipse each other. This cyclical dimming and brightening unveils the stars' sizes and orbital periods, providing a critical key to unlocking their secrets.

Stars come in a stunning variety, and binary systems offer a captivating mix of stellar combinations. Sun-like stars orbiting at vast distances, like Alpha Centauri A and B, are one example, providing a relatively stable environment for potential planetary systems. However, close binary systems, where stars are gravitationally locked in a tight embrace, hold a special intrigue. In some cases, these stars are so close that their atmospheres intertwine, exchanging material in a cosmic game of stellar tag. Extreme proximity can even lead to direct contact, creating a dazzling spectacle of two massive, hot stars practically merging.

The drama intensifies when one star is a compact object – a white dwarf, a neutron star, or even a black hole. The immense gravitational pull of the compact object exerts a vampiric influence, stealing material from its stellar companion. If the compact object is a white dwarf feeding on gas from a companion, the result is a cataclysmic variable star, where the inflowing gas ignites and emits powerful radiation. These stellar vampires are aptly nicknamed "cataclysmic variables," and their outbursts can be spectacular, briefly outshining their companion stars.

When a neutron star or black hole is the culprit, the system is called an X-ray binary. The category is further divided into low-mass or high-mass X-ray binaries depending on the size of the donor star, the one feeding the compact object. These systems can be tremendously energetic, as the infalling material is heated to scorching temperatures by the immense gravity of the compact object, releasing a torrent of X-rays.

A captivating example is AR Scorpii, a binary pulsar system. Here, a white dwarf pulsar, about the size of Earth, waltzes with a red dwarf companion. Pulsars are rapidly spinning, highly magnetized objects that emit powerful beams of radiation in a precise, periodic manner. While typically neutron stars, pulsars can also be white dwarfs, although their slower rotation betrays their less dense nature. The discovery of AR Scorpii, the first such system identified, forced astronomers to rewrite the textbooks on stellar evolution in binary systems.

Binary systems defy simplicity, their evolution a captivating cosmic drama. Consider a binary with two hefty main sequence stars, each boasting a mass exceeding our Sun's. These stars burn brightly and fast, consuming their hydrogen fuel at an alarming rate. Inevitably, one star expands beyond its Roche lobe, a boundary marking the gravitational influence of each star. This expansion triggers mass transfer, with one star siphoning material from its companion. The stolen bounty can form an accretion disk, a swirling vortex of stellar material feeding the recipient star, or it can be directly absorbed through collisions.

The vampiric star spins faster, its shape warping under the influence of the stolen mass. Meanwhile, the donor star sheds a significant portion of its mass, potentially shrinking to a fraction of its original size. The recipient star, now a heavyweight, undergoes a dramatic increase in its fusion rate. This stellar gluttony fuels a powerful stellar wind that further diminishes the donor star. The once hefty companion withers, destined to explode as a supernova, leaving behind a tiny neutron star remnant that might even escape the system altogether. The vampiric star, having gorged itself, inflates into a red supergiant before its own supernova demise, also likely leaving behind a neutron star. Binary systems showcase stellar evolution on a fast track, each star's fate intricately woven to the other, creating a complex tapestry of gravitational interactions, mass transfer, and explosive events.

The universe abounds with fascinating stellar configurations – binary, triple, and even more complex systems. Remarkably, some of these celestial dance partners harbor planets. Imagine a world bathed in the light of two, three, or even more suns, their orbits casting an ever-changing light show on the alien landscape below. The day-night cycle would be a complex dance, with periods of blazing light alternating with intervals of twilight or even complete darkness. The gravitational influence of these multiple suns would the gravitational influence of these multiple suns would undoubtedly affect the orbital mechanics of any planets within the system, potentially leading to more eccentric orbits or even chaotic interactions. However, such complexity could also foster unique conditions, creating environments that might push the boundaries of what we traditionally consider habitable. Future astronomical discoveries may soon unveil the secrets these intriguing systems hold, offering a glimpse into the remarkable diversity of planetary environments and the potential for life beyond our wildest imaginations.

THE COSMIC CHRONOLOGY

Chapter 48
General Theory of Relativity (corroborated)

Albert Einstein's theory of general relativity revolutionized our understanding of gravity. This theory proposes that gravity isn't a force, but rather a warping of spacetime, the four-dimensional fabric that weaves together space and time. While visualizing this warping in three dimensions is challenging, let alone on a flat screen, it can be likened to a heavy object placed on a trampoline, creating a depression that alters the path of a marble rolling across the surface.

General relativity, published in 1915, received its first confirmation just four years later. Einstein predicted that light, traveling at incredible speeds, would bend around massive objects due to the curvature of spacetime. This theory was validated during a solar eclipse expedition led by Arthur Eddington, where starlight was observed to be slightly deflected by the Sun's gravity, precisely as Einstein predicted.

This marked the beginning of a series of victories for general relativity. Black holes, theorized earlier, emerged as prime examples of the theory's power. These collapsed stars, with their immense density and warped spacetime, trap even light within their grasp. Technology eventually caught up, allowing us to observe the supermassive black hole residing at the heart of our Milky Way galaxy, a feature likely common to most galaxies.

Beyond direct black hole observation, studying the motion of stars near the galactic center strengthens the case for general relativity. These stars defy explanation without the presence of the supermassive black hole. Their orbits trace paths around a seemingly empty point, accelerating to phenomenal speeds as they approach the black hole before slingshotting away in a cosmic ballet. Furthermore, these stars exhibit a precession in their orbits, similar to Mercury's precession around our Sun – a phenomenon solely predicted by the equations of general relativity.

The reach of general relativity extends beyond our own galaxy. Powerful telescopes like Hubble have unveiled a universe teeming with distant galaxies. One captivating prediction of the theory, gravitational lensing, is regularly observed. This phenomenon occurs when the light from a distant object is bent by the gravity of another object, like a closer galaxy, creating a distorted and magnified view. This cosmic lensing offers valuable insights into both the lensed object and the lensing object, akin to the 1919 Eddington observation but on a grander, intergalactic scale.

In certain gravitational lensing scenarios, the light bends around the foreground object in a complete circle, forming an "Einstein ring" – a breathtaking display of light. General relativity, treating spacetime as a fabric, even predicts ripples in this fabric caused by the motion of massive objects. These ripples, called gravitational waves, are notoriously difficult to detect due to the weakness of gravity. However, in 2016, LIGO and other gravitational wave observatories achieved the seemingly impossible – directly detecting gravitational waves from cataclysmic events like black hole mergers and neutron star mergers.

The incredible density of neutron stars allows the gravitational waves generated by their mergers to reach Earth, where they can be picked up by highly sensitive equipment like LIGO's intricate network of lasers and mirrors. These detections not only validate general relativity's predictions but also unveil the spectacular "kilonova" explosions that follow such mergers. These explosions are thought to be the primary source of heavy elements in the universe, like gold and platinum.

From supermassive black holes and their orbiting stars to gravitational lensing and the elusive gravitational waves, the evidence for general relativity is overwhelming. The theory's equations predict the existence and properties of these phenomena with remarkable accuracy, and observations over the past decades have confirmed these predictions in exquisite detail. General relativity's influence extends beyond these phenomena. For instance, the precise functioning of GPS satellites relies on corrections for gravitational time dilation, another facet of the theory.

Over a century since its inception, general relativity remains the reigning champion of gravity. Challenges remain, however. Reconciling general relativity with quantum mechanics remains a hurdle, particularly when describing phenomena like singularities and the universe's earliest moments. But that, as they say, is a story for another day, perhaps waiting to be unraveled by another visionary mind like Einstein.

Chapter 41
Quasars and the Older Light

Gazing into the cosmos is akin to peering into the past. The vast distances between celestial objects translate to looking back in time as we train our telescopes on their faint light. Objects billions of light-years away reveal the universe's youthful state, offering a window into an era when the first large-scale structures were beginning to take shape.

This fortunate circumstance allows us to delve into the enigmatic realm of quasars, incredibly luminous celestial bodies from the early universe. When these faint, star-like objects were first observed in the 1950s, they were a complete mystery, leading to the moniker "quasi-stellar object" (or quasar for short). However, a closer look through the lens of spectroscopy revealed a startling truth – these objects weren't faint stars at all, but rather extremely distant, active galactic nuclei (AGNs), blazing beacons outshining entire galaxies in their brilliance.

At the heart of a quasar lies a supermassive black hole, dwarfing our Sun by millions or even billions of times. Surrounding this monstrous entity is a swirling disk of gas, acting as its fuel source. As this gas spirals inwards towards the black hole's immense gravity, incredible friction generates tremendous heat. This fiery process makes the quasar glow with an intensity thousands of times greater than an entire galaxy like our Milky Way.

The immense luminosity of quasars, despite their vast distances, initially baffled astronomers. However, over time, a deeper understanding emerged. We now recognize quasars as the seeds for future galaxies. Their immense gravitational pull acts like a cosmic magnet, drawing in gas and fueling their brilliant glow. This very process lays the groundwork for galaxy formation, with the quasar acting as a precursor to the majestic spiral arms and vibrant nebulae we observe in mature galaxies.

Most quasars are estimated to have formed around ten billion years ago and reside within host galaxies. These galactic companions provide further evidence for the connection between quasars and galaxy formation. Interestingly, only one quasar lacking a host galaxy has been found so far. Studying such anomalies could be key to unlocking the secrets of galaxy formation. Some intriguing theories propose that quasar jets, powerful streams of particles ejected from the supermassive black hole, might trigger galaxy development. These nascent galaxies could eventually merge with the quasar itself, creating even more massive structures within the universe.

The furthest quasars we have observed push the boundaries of cosmic time. Some of these celestial giants are thought to have formed a mere 690 million years after the Big Bang, during the universe's "dark ages" – a period shrouded in relative obscurity. These primeval quasars often reside within giant gas halos, vast reservoirs of cool hydrogen gas fueling the supermassive black hole's growth and contributing to the quasar's immense luminosity.

Unraveling the mysteries of quasars involved a multifaceted approach, utilizing multiple lines of evidence. X-ray astronomy, spectral analysis, and gravitational lensing all played crucial roles. Gravitational lensing, for instance, can distort a quasar's light from a foreground galaxy, creating a spectacular phenomenon known as an Einstein Cross – where the quasar appears as four distinct images.

Modern astronomy takes this analysis even further by distinguishing between macro and microlensing. Macrolensing, involving a foreground galaxy, acts as a natural magnifying glass, allowing us to observe these incredibly distant objects. Microlensing, caused by individual stars within the foreground galaxy, provides even more detailed information. By studying the subtle variations in the quasar's brightness caused by microlensing, astronomers can glean insights into the properties of the foreground stars and further refine our understanding of the quasar itself.

At the forefront of quasar research lies a powerful technique called Very Long Baseline Interferometry (VLBI). This technique essentially links multiple telescopes worldwide to act as a single, giant telescope. This collaborative effort allows for incredibly sharp observations of these distant objects, providing us with unprecedented detail about their structure and behavior.

Current research on quasars delves deeper, focusing on the large-scale distribution of these celestial bodies across billions of light-years within cosmic structures. Additionally, astronomers are investigating the alignment between the spin axes of quasars and the structures they reside in. These lines of inquiry promise to unlock new secrets about the universe's formation and evolution. By studying quasars, we gain a deeper appreciation for the grand narrative of the cosmos, from the fiery birth of these active galactic nuclei to their potential role as the seeds for majestic galaxies. As we continue to unravel the mysteries of quasars, we take another step forward in our quest to understand the ever-expanding universe. Quasars, with their immense luminosity piercing the veil of the Dark Ages, offer a powerful tool to probe the universe's earliest moments. Studying these primeval objects can illuminate the formation of the first stars and galaxies, a period shrouded in mystery. By analyzing the distribution of quasars across vast cosmic distances, we can map the intricate network of filaments, voids, and galaxy clusters that form the large-scale structure of the universe – the cosmic scaffolding upon which our own Milky Way resides.

These celestial giants also provide a unique window into the feeding frenzy of supermassive black holes at their heart. By studying the properties of quasar accretion disks and jets, astronomers can gain valuable insights into how these black holes grow and evolve over time. The extreme environments within quasars act as a natural laboratory, pushing the limits of our understanding of gravity, particularly Einstein's theory of general relativity. Observing the behavior of light and matter under the influence of these supermassive black holes can reveal potential discrepancies or necessitate new theories altogether.

Furthermore, the vast luminosities of quasars can be used to probe the elusive dark matter and dark energy that dominate the universe's composition. Studying how quasars interact with these invisible entities might provide clues to their nature and role in cosmic evolution. The quest to unravel the mysteries of quasars is a continuous journey, each discovery opening doors to further exploration. As we refine our observational techniques and delve deeper into the secrets these celestial giants hold, we gain a more comprehensive understanding of the grand narrative of the cosmos, from the faint echoes of the Big Bang to the ongoing dance of galaxies across the vast expanse of space and time.

THE COSMIC CHRONOLOGY

Chapter 42
Scaling the observable universe

The universe. We all know it's vast, but grasping its true scale can be mind-boggling. To wrap our heads around it, let's embark on a journey, starting from our familiar home, Earth.

Earth, with its 8,000-mile diameter, seems massive to us. But in the cosmic scheme, it's a mere speck. Even our closest celestial neighbor, the Moon, is a staggering 384,000 kilometers away – over 30 Earths lined up in a row!

Zooming out, we encounter the inner solar system, encompassing the Sun and the four rocky planets. This stretches to about 456 million kilometers, over a thousand times the Earth-Moon distance. But that's just the beginning.

Extending further out is the outer solar system, reaching Neptune and dwarf planets. Here, distances balloon to a mind-numbing 9 billion kilometers – roughly 20 times the diameter of the inner solar system.

Leaving our solar system, we enter the realm of stars. Here, kilometers become inadequate. We switch to light-years – the distance light travels in a year, a whopping 9.4 trillion kilometers. Our nearest stellar neighbor, Proxima Centauri, lies a staggering 4 light-years away. Imagine lining up thousands of solar systems just to reach it!

Interstellar space is vast and empty. Don't be fooled by artistic depictions of closely packed stars. Our local stellar neighborhood, with just a handful of stars, spans a vast 30-40 light-years.

Now, let's truly zoom out. Our Milky Way galaxy, with its 100,000-light-year diameter, dwarfs everything we've seen so far. The stars we see at night are mere residents in one of its spiral arms, our cosmic home.

The Milky Way isn't alone. It resides in the Local Group, a collection of around 30 galaxies, with Andromeda, our soon-to-be collision partner, being the second largest. This group stretches 10 million light-years across.

Superclusters are the next level up. Laniakea, our supercluster, houses a staggering 100,000 galaxies, including our Virgo supercluster, a smaller region within Laniakea with a diameter of 500 million light-years.

Even vaster structures exist – systems of adjacent superclusters containing millions of galaxies, spanning at least 3 billion light-years. At this scale, individual galaxies become indistinguishable, resembling faint dust clouds.

Finally, we reach the observable universe, a mind-blowing 93 billion light-years across. This limit is set by the universe's age – light hasn't had more time to travel and reach us. The true size of the entire universe? We simply don't know, but estimates suggest it could be 250 times larger than the observable universe!

Our journey has traversed vast distances in leaps, showcasing the universe's hierarchical structure. But to truly appreciate its scale, let's embark on a continuous voyage, from Earth's surface to the very edge of the observable universe and back. This mind-bending odyssey awaits... Fasten your metaphorical seatbelts, for we're about to embark on a mind-bending odyssey that defies the limitations of physics. This continuous journey will take us from the familiar ground beneath our feet to the very edge of the observable universe, and then all the way back. Prepare to be humbled by the sheer immensity of the cosmos.

Leaving Earth: As we lift off, Earth shrinks into a vibrant blue marble suspended in the blackness. Within seconds, continents blur into swirling patterns, and soon, even our entire planet becomes a mere speck against the vast canvas of space.

Solar System: Within minutes, the Moon dwindles to a distant, grayish orb. The familiar planets – Mercury, Venus, Mars – become fleeting points of light as we hurtle past them at an unimaginable speed. Even the mighty Sun, the source of life on Earth, shrinks to a shrinking yellow dot in our rearview mirror.

Interstellar Space: Hours turn into days as we traverse the desolate expanse of interstellar space. Here, the only companions are the occasional wispy nebula or the faint glow of a distant star. The emptiness is both awe-inspiring and isolating.

A Crossing of Light-Years: Weeks, months, even years pass in a blur of cosmic nothingness. We cross the threshold of the first light-year – a distance so immense that it boggles the mind. Yet, this is just a drop in the cosmic ocean. Proxima Centauri, our nearest stellar neighbor, remains stubbornly distant.

The Milky Way Galaxy: Finally, after years of relentless travel, we breach the edge of the solar system and enter the swirling embrace of the Milky Way galaxy. We hurtle past countless stars, each a potential sun with its own planetary system. The sheer number of celestial bodies is overwhelming, a testament to the galaxy's vastness.

Galactic Traverse: Decades, perhaps centuries, tick by as we race through the Milky Way's spiral arms. We witness the birth and death of stars, the formation of star clusters, and the majestic dance of nebulae. The galaxy's immense scale becomes a tangible reality.

Beyond the Milky Way: Centuries turn into millennia as we finally reach the Milky Way's outer reaches. Here, the density of stars thins, and the vastness of intergalactic space unfolds once again. We encounter other galaxies, each a unique island in the cosmic sea. Some are majestic spirals like our own, while others are irregular clumps of stars.

The Local Group and Beyond: Millennia bleed into eons as we zip past the neighboring galaxies of the Local Group. Andromeda, our galactic counterpart, looms large, a silent reminder of the eventual collision that awaits our Milky Way.

Supercluster Spectacle: As eons stretch into unimaginable stretches of time, we enter the realm of superclusters. Laniakea, our home supercluster, unfolds before us, a colossal structure teeming with hundreds of thousands of galaxies. The Virgo Supercluster, a smaller region within Laniakea, becomes a fleeting landmark on our cosmic voyage.

The Edge of the Observable Universe: Finally, after countless eons, we reach the edge of the observable universe. Here, the cosmic microwave background radiation, a faint echo of the Big Bang, intensifies. We have reached the limit of what we can see – the farthest point from which light has had enough time to reach us.

The Journey Back: But our odyssey isn't over. We reverse course, retracing our steps through the eons. The superclusters, galaxies, and the vast emptiness of interstellar space rush by in reverse order. The Milky Way regains its dominance, then the solar system, and finally, Earth comes into view once more.

Landing on Earth: After an incomprehensible journey through time and space, we touch down on the familiar ground.

The universe appears unchanged, yet our perspective has been irrevocably altered. We have witnessed the cosmos on a scale beyond human comprehension, a humbling reminder of our place in the vast expanse of existence.

This mind-bending voyage has hopefully instilled a newfound appreciation for the universe's staggering immensity. The vastness of space, the dance of galaxies, and the sheer scale of time all contribute to the awe-inspiring beauty of the cosmos. As we continue to explore and learn, the universe will continue to unveil its secrets, forever pushing the boundaries of our understanding.

THE COSMIC CHRONOLOGY

Chapter 43
Tribute to World's Geniuses and Machines

Isaac Newton (1643-1727): The Architect of Classical Physics

Isaac Newton stands as a giant upon whose shoulders countless scientists have built. His contributions to astronomy and mathematics are nothing short of revolutionary. Newton's laws of motion, formulated in the 17th century, laid the groundwork for classical mechanics. These laws – the law of inertia, the law of acceleration, and the law of action and reaction – provided a universal framework for understanding how objects move and interact with forces.

In his seminal work, "Philosophiæ Naturalis Principia Mathematica" (Mathematical Principles of Natural Philosophy), Newton not only explained terrestrial motion but also celestial motion. He demonstrated that the same force that causes an apple to fall to the ground also governs the movement of planets around the Sun. This unifying concept, universal gravitation, explained Kepler's laws of planetary motion with remarkable accuracy.

Newton's law of universal gravitation posits that every particle in the universe attracts every other particle with a force proportional to the product of their masses and inversely proportional to the square of the distance between them. This simple yet powerful law could explain the elliptical orbits of planets, the motion of moons around planets, and even the tides on Earth.

Newton's mathematical genius further shone through his development of calculus, an indispensable tool for solving complex problems in physics, engineering, and other scientific disciplines. Calculus provided the mathematical language to describe and analyze motion, change, and rates of change – concepts that are fundamental to our understanding of the universe.

Galileo Galilei (1564-1642): The Champion of Observation

Galileo Galilei, often hailed as the "father of observational astronomy," revolutionized our view of the cosmos through his pioneering use of the telescope. Before Galileo, astronomy relied heavily on the naked eye and rudimentary instruments. In 1609, he turned his homemade telescope towards the night sky and made a series of groundbreaking discoveries.

Galileo observed mountains and craters on the Moon, shattering the long-held belief of a smooth lunar surface. He discovered four moons orbiting Jupiter, demonstrating that not everything revolved around the Earth. He observed the phases of Venus, providing strong evidence that Venus, like Earth, orbited the Sun. These observations directly contradicted the geocentric model and provided crucial support for the heliocentric model.

Galileo's championing of the heliocentric model and his outspoken nature put him at odds with the Catholic Church, which adhered to the geocentric view. He was forced to recant his beliefs under threat of torture, but his observations and scientific contributions left an indelible mark on astronomy.

Srinivasa Ramanujan (1887-1920): The Mathematical Mystic

Srinivasa Ramanujan, an Indian mathematician with little formal education, possessed an extraordinary mathematical genius. Though largely self-taught, he produced groundbreaking work in number theory, infinite series, and mathematical analysis. While his contributions to astronomy were not direct, his mathematical insights have had profound implications for our understanding of the universe.

Ramanujan's work on highly composite numbers, for example, has applications in areas like black hole physics. Black holes are regions in spacetime with such intense gravity that nothing, not even light, can escape. Studying the properties of highly composite numbers helps us understand the behavior of matter under extreme gravitational conditions.

His work on mock theta functions, a complex area of mathematics, has potential applications in cosmology, the study of the large-scale structure and evolution of the universe. These functions may help us understand the distribution of galaxies and dark matter in the cosmos.

Ramanujan's brilliance and his ability to make profound discoveries with limited resources continue to inspire mathematicians and scientists worldwide.

Vikram Sarabhai (1919-1971): The Father of the Indian Space Program

Vikram Sarabhai, a visionary scientist and entrepreneur, is widely regarded as the father of the Indian space program. He played a pivotal role in establishing India's space research capabilities and laying the foundation for its future endeavors. Recognizing the importance of space technology for development and national security, Sarabhai actively promoted space research in India.

He established the Physical Research Laboratory (PRL) in Ahmedabad in 1947, which became a crucial center for space science research. He founded the Indian National Committee for Space Research (INCOSPAR) in 1962, which later evolved into the Indian Space Research Organisation (ISRO). Under his leadership, India launched its first sounding rocket in 1963, marking a significant milestone in the nation's space journey.

Sarabhai's vision extended beyond scientific exploration. He believed that space technology could be harnessed for practical applications like telecommunications, weather forecasting, and resource management. He actively pursued the development of satellite-based communication systems, laying the groundwork for India's robust telecommunications infrastructure today.

A.P.J. Abdul Kalam (1931-2015): From Missiles to the Moon

A.P.J. Abdul Kalam, a renowned aerospace scientist and former president of India, earned the moniker "Missile Man of India" for his pivotal role in developing the nation's missile program. His contributions significantly strengthened India's defense capabilities and technological prowess.

Kalam joined India's space program in the 1960s and played a key role in the development of the SLV-3, India's first satellite launch vehicle. He later led the project to develop Agni, a series of ballistic missiles that have become a cornerstone of India's defense system.

Kalam's vision extended beyond military applications. He recognized the potential of missile technology for civilian purposes and championed the development of launch vehicles for space exploration. His contributions paved the way for India's successful Chandrayaan-1 mission, which sent a lunar probe to the Moon in 2008.

Nambi Narayanan (born 1942): The Unsung Hero of Cryogenic Propulsion

Nambi Narayanan, a distinguished aerospace engineer, has made significant contributions to the development of India's cryogenic engine technology. Cryogenic engines, which burn fuel at extremely low temperatures, are far more efficient than conventional rocket engines and are crucial for launching heavier payloads into space.

Narayanan played a key role in the development of the Vikas engine, which powers India's Polar Satellite Launch Vehicle (PSLV). He also led the cryogenic engine development program, facing significant challenges and overcoming political hurdles.

Despite these challenges, Narayanan's dedication and expertise have made India self-sufficient in cryogenic technology, a critical milestone in the nation's space program. He continues to inspire future generations of scientists and engineers working in the field of aerospace propulsion.

Ritu Karidhal (born 1978): A Rising Star in Space Exploration

Ritu Karidhal, a young and accomplished astrophysicist, is making significant strides in our understanding of the formation and evolution of galaxies. Her research focuses on the interaction between galaxies, particularly how mergers and collisions between galaxies can trigger star formation, influence galactic structure, and potentially lead to the formation of supermassive black holes.

Karidhal's expertise lies in mission planning, spacecraft operations, and data analysis. She played a crucial role in India's Mars Orbiter Mission (Mangalyaan), serving as the Deputy Operations Director. Her meticulous planning and leadership were instrumental in the successful insertion of the spacecraft into Mars' orbit, making India the first Asian country and the fourth space agency in the world to achieve this feat.

Currently, Karidhal is the Mission Director for Chandrayaan-3, India's ambitious lunar mission aimed at a soft landing on the lunar surface. Her dedication and expertise are propelling India's space exploration endeavors to new heights, and she serves as an inspiration for young women aspiring to careers in science and technology.

Kalpana Chawla (1962-2003): Soaring High, Inspiring Millions

Kalpana Chawla, an Indian-born American astronaut and aerospace engineer, blazed a trail as the first woman of Indian origin to fly into space. Her life and career embodied courage, dedication, and a passion for exploration.

Born in Karnal, India, Chawla developed a fascination with airplanes and flight from a young age. She pursued her dream, earning a Bachelor of Engineering degree in Aeronautical Engineering from Punjab Engineering College in India. Driven by her ambition, she moved to the United States, where she obtained a Master's degree and a Ph.D. in Aerospace Engineering.

Chawla's expertise in aerospace engineering landed her a coveted position at NASA. Selected as a member of the astronaut corps in 1994, she underwent rigorous training to prepare for space missions. In 1997, she made history as a mission specialist and the primary robotic arm operator on the Space Shuttle Columbia's STS-87 mission.

During this mission, Chawla played a critical role in deploying the Spartan satellite, a scientific research platform. She logged over 31 days in space, conducting experiments and contributing significantly to the mission's success. Chawla's historic flight ignited a spark of inspiration in millions, particularly young girls in India and around the world. She became a symbol of achievement, demonstrating that with dedication and hard work, one can reach for the stars.

Tragically, Chawla perished along with six other crew members in the Columbia disaster on February 1, 2003. The Space Shuttle disintegrated upon re-entry into Earth's atmosphere. While the loss was deeply felt by the scientific community and the world, Kalpana Chawla's legacy continues to inspire generations. Her story exemplifies the pursuit of scientific exploration, the importance of diversity in STEM fields, and the enduring human spirit that strives to reach beyond the known.

Machines:

Sputnik 1

Sputnik 1, launched by the Soviet Union in 1957, holds the distinction of being the first artificial satellite to orbit Earth. This event marked the dawn of the Space Age, igniting a global fascination with space exploration. The beeping satellite, a simple metal sphere, orbited Earth for three weeks before falling back into the atmosphere. Sputnik 1's success ushered in a new era of scientific discovery and competition, particularly between the US and the Soviet Union, known as the Space Race.

Pioneer 10

Pioneer 10, launched in 1972, was a robotic spacecraft on a pioneering mission to explore the outer reaches of our solar system. It became the first human-made object to fly past Jupiter in 1973 and escape the solar system's gravity altogether. Pioneer 10 carried a plaque with a message and images depicting humanity and our place in the cosmos, intended for any potential extraterrestrial civilizations that might encounter it in the vast interstellar space. The spacecraft continues its journey even today, though communication with it ceased in 2000.

Voyager 1 & 2

Voyager 1 and 2, launched in 1977, are twin robotic spacecraft on an epic voyage of exploration that has taken them far beyond the outer planets of our solar system. Voyager 1, despite being launched later, took a faster trajectory and became the first spacecraft to enter interstellar space in 2012. Voyager 2 followed suit in 2018. Both spacecraft carry a

Golden Record, a collection of sounds, images, and messages intended for any alien civilizations that might find them, offering a glimpse of humanity and Earth. They continue to transmit valuable scientific data back to Earth, providing insights into the mysterious realm of interstellar space.

Hubble Space Telescope

The Hubble Space Telescope, launched in 1990, revolutionized our understanding of the universe. Unlike previous telescopes limited by Earth's atmosphere, Hubble observes from space, offering a clear view of the cosmos. It has captured breathtaking images of distant galaxies, nebulae, and stars, providing astronomers with invaluable data on the formation and evolution of the universe. Hubble's discoveries have challenged our existing theories and opened up new avenues for astronomical research, forever changing our perspective on our place in the cosmos.

Cassini

Cassini, a robotic spacecraft launched in 1997, spent over a decade orbiting Saturn, its rings, and moons. It provided us with the most detailed observations of this complex planetary system ever obtained. Cassini made groundbreaking discoveries, revealing the beauty and complexity of Saturn's rings, the active atmosphere of Titan, and the vast oceans beneath the icy surface of Enceladus. The mission concluded in a dramatic plunge into Saturn's atmosphere in 2017, but the data it collected continues to be a treasure trove for scientists studying Saturn and the possibility of life elsewhere in our solar system.

Curiosity

Curiosity, a rover launched by NASA in 2011, has been exploring the surface of Mars since 2012. This six-wheeled marvel is equipped with a suite of scientific instruments designed to search for signs of past or present life on the Red Planet. It has analyzed Martian rocks and soil, drilled deep beneath the surface, and even taken selfies amidst the alien landscape. Curiosity's discoveries have helped us understand the Martian environment, revealing evidence of ancient lakes and a potentially habitable past. The rover continues its mission, venturing further into the Martian terrain and unraveling the secrets this enigmatic planet holds.

Insight

Insight, a robotic lander that touched down on Mars in 2018, is on a mission to study the planet's interior. Unlike rovers that explore the surface, Insight delves deep, using sophisticated instruments to measure Mars' seismic activity, heat flow, and internal structure. This data will shed light on how Mars formed and evolved, helping us understand the processes that shaped our rocky neighbor. By studying Mars' interior, Insight can also help us determine if the planet ever had an active magnetosphere and potentially supported life in the past.

Deep Space

Deep space refers to the vast and largely unexplored region beyond our solar system. It encompasses everything from interstellar space, the region between stars, to distant galaxies and intergalactic medium. This realm holds countless mysteries, from the nature of dark matter and dark energy to the potential existence of other planetary systems and life beyond Earth. Exploring deep space is a challenging endeavor, but missions like Voyager 1 & 2 and future endeavors like the

James Webb Space Telescope are pushing the boundaries of our knowledge and helping us unravel the secrets of the cosmos.

Mangalyaan (Mars Orbiter Mission)

Mangalyaan, also known as the Mars Orbiter Mission (MOM), launched by India in 2013, successfully entered Martian orbit in 2014. This remarkable mission marked India's foray into interplanetary exploration and made it the fourth space agency to achieve Martian orbit. MOM carries five scientific instruments designed to study the Martian atmosphere, composition, and surface features. Its success not only boosted India's spacefaring capabilities but also demonstrated the feasibility of cost-effective Mars missions, opening doors for further exploration.

Chandrayaan 1 & 2

Chandrayaan 1 and 2 are India's lunar exploration missions. Chandrayaan 1, launched in 2008, was India's first mission to the Moon. It successfully orbited the Moon and carried out scientific studies for several months. Notably, Chandrayaan 1 confirmed the presence of water ice on the lunar surface, a significant discovery that has implications for future lunar exploration.

Chandrayaan 2, launched in 2019, aimed for a more ambitious landing mission. It consisted of a lunar orbiter, a lander named Vikram, and a rover named Pragyaan. While the orbiter successfully entered lunar orbit and continues to send back valuable data, the Vikram lander experienced a hard landing and communication was lost. Despite this setback, Chandrayaan 2 marked a significant advancement in India's lunar exploration capabilities and paves the way for future missions.

Chandrayaan 3

Chandrayaan 3 is the next chapter in India's lunar exploration story. Planned for a future launch, it is designed to be a lander mission and will attempt a soft landing on the lunar surface. Unlike Chandrayaan 2, it will not deploy a rover. The primary objective of Chandrayaan 3 is to demonstrate India's ability to perform a soft landing on the Moon and further our understanding of the lunar surface composition. The successful completion of this mission will be a major milestone for India's space program.

Aditya-L1

Aditya-L1 is an upcoming Indian mission dedicated to studying the Sun. Scheduled for launch in 2024, it will be placed in a halo orbit around the L1 Lagrange point, a point where the gravitational forces of the Sun and Earth balance each other. This vantage point will allow Aditya-L1 to continuously observe the Sun's corona, the outermost layer of its atmosphere, which plays a crucial role in space weather events that can impact Earth's communication and power grids. By studying the Sun's corona, Aditya-L1 will provide valuable data for space weather forecasting and help us understand the Sun's influence on Earth's environment.

Aryabhatta and Rohini

Aryabhatta and Rohini are significant satellites in the history of India's space program. Aryabhatta, launched in 1975, was India's first indigenously built satellite. It was an experimental satellite designed to test

India's satellite building capabilities in areas like communication and remote sensing. The success of Aryabhatta paved the way for future Indian satellites.

Rohini satellites, a series launched from the late 1970s onwards, were primarily used for communication and remote sensing purposes. They played a vital role in establishing India's communication infrastructure and providing valuable data for resource management and disaster monitoring. The Rohini series helped India gain experience in satellite design, operation, and applications, laying the foundation for more advanced satellites in the future.

James Webb Space Telescope

The James Webb Space Telescope (JWST), launched in December 2021, is a marvel of modern engineering and a testament to international collaboration. Designed primarily for infrared astronomy, it observes the universe in a wider range of wavelengths than its predecessor, the Hubble Space Telescope. This allows JWST to peer further back in time, potentially revealing the formation of the first stars and galaxies just a few hundred million years after the Big Bang.

JWST's high-resolution and high-sensitivity instruments enable it to observe objects that are too old, distant, or faint for Hubble to detect. It can study the atmospheres of exoplanets, planets outside our solar system, searching for potential biosignatures, molecules that might indicate the possibility of life. Additionally, JWST can investigate the formation of stars and planetary systems within our own galaxy and shed light on the mysterious processes that govern the universe's evolution.

The James Webb Space Telescope is a revolutionary tool for astronomers, promising to revolutionize our understanding of the cosmos. With its ongoing observations, we can expect groundbreaking discoveries that will reshape our perspective on the universe's origins, our place within it, and the potential for life beyond Earth.

"Tribute everyone who has contributed to this world through the stream of astronomy, cosmology and astrophysics and many more..."

Chapter 44
Top Space research Organizations of the World

National Aeronautics and Space Administration (NASA):

Undoubtedly the most recognizable name in space exploration, NASA is the civilian space agency of the United States. Established in 1958, it played a pivotal role in the Space Race against the Soviet Union and has been instrumental in numerous groundbreaking achievements. From the Apollo missions that landed humans on the Moon to the Voyager probes venturing into interstellar space, NASA's contributions to space science and exploration are unparalleled. It continues to be at the forefront of space exploration, leading missions to Mars, studying the outer solar system, and developing innovative technologies to push the boundaries of human knowledge.

China National Space Administration (CNSA):

A relative newcomer compared to NASA, CNSA has rapidly emerged as a major player in space exploration. Founded in 1993, it has made significant strides in recent years. CNSA has successfully landed robotic rovers on the Moon, launched its own space station, Tianhe, and aims for crewed lunar missions in the future. They are also actively involved in planetary exploration with missions to Mars. CNSA's ambitious goals and rapid advancements make it a force to be reckoned with in the future of space exploration.

European Space Agency (ESA):

A collaborative effort of 22 member states, ESA is a unique space agency that fosters cooperation among European nations. Established in 1975, it has played a crucial role in developing Europe's space capabilities. ESA has made significant contributions to space science through its probes and missions, like the Rosetta mission that landed a probe on a comet. They partner with other space agencies on major projects like the International Space Station and develop launch vehicles like Ariane rockets. ESA's collaborative approach and focus on scientific discovery make it a vital player in the global spacefaring community.

Japan Aerospace Exploration Agency (JAXA):

Founded in 2003 through the merger of existing space agencies, JAXA is responsible for Japan's space program. They have a strong focus on robotic exploration, having deployed probes to the Moon and partnered with other space agencies on missions like the International Space Station and asteroid exploration missions. JAXA is also developing its own reusable launch vehicle and aims for future crewed missions to the Moon. Their dedication to technological innovation and scientific exploration makes JAXA a key player in the global space race.

Indian Space Research Organisation (ISRO):

Established in 1969, ISRO is the space agency of India. Renowned for its cost-effective missions, ISRO has made remarkable achievements in recent years. They successfully launched the Mangalyaan orbiter to Mars, making India the fourth space agency to achieve Martian orbit. ISRO has also undertaken lunar exploration missions with Chandrayaan-1 and the attempt to land Vikram lander and deploy Pragyaan rover with Chandrayaan-2. ISRO's focus on developing its own launch vehicles and technologies, coupled with its ambitious missions, makes it a rapidly growing space agency on the world stage.

Chapter 45
The End of the Universe

Our universe stretches vast and unimaginable, a cosmic tapestry woven with the luminous threads of galaxies, the twinkling points of stars, and the invisible tendrils of dark matter and energy. Yet, even the grandest narratives must reach a conclusion, and our cosmos is no exception. Today, we embark on a thought-provoking exploration to consider the potential fates awaiting everything we know.

One possibility paints a fiery picture – the Big Crunch. Imagine the universe as a colossal balloon, perpetually inflating. In this scenario, the expansion would eventually slow, grind to a halt, and then reverse. Everything, from the farthest galaxies to the tiniest subatomic particles, would hurtle back towards a single point, potentially igniting another Big Bang and restarting the cosmic cycle. This "Big Bounce" universe throws open a Pandora's box of mind-bending questions. Would each iteration be an identical replica, or would the fundamental constants governing physics be rewritten? Did this very conversation occur a trillion years ago, destined to repeat for eternity in an endless loop of cosmic existence?

Alternatively, the universe could face a far frostier fate. Here, the expansion would continue forever, driven by a mysterious force known as dark energy. Galaxies would become increasingly isolated lighthouses in the ever-expanding cosmos, stretching further and further apart until they vanish from each other's view. Star formation, once a vibrant process birthing countless suns, would cease as the once-abundant fuel dwindled. Over eons that stagger the human imagination, even the sturdiest atoms would succumb to decay, their protons dissolving into their constituent quarks. Black holes, those monstrous consumers of matter and energy, would evaporate, their immense gravity no match for the relentless expansion. Absolute zero would reign supreme, leaving only an empty, ever-expanding void – the chilling signature of the Big Rip.

These scenarios are based on our current understanding, particularly our limited knowledge of dark energy. What if this enigmatic force isn't constant? Perhaps it weakens over time, allowing gravity to reassert its dominance and usher in the Big Crunch. Or maybe dark energy harbors some unforeseen properties that completely alter the cosmic equation. New discoveries, lurking just beyond the horizon of our knowledge, could radically transform our predictions.

Before the discovery of dark energy, cosmologists grappled with the fundamental shape of the universe. Was it open, destined to forever expand in an ever-chilling embrace? Closed, fated to collapse in on itself in a fiery inferno? Or flat, delicately balanced between these two extremes? To answer this question, we meticulously measured the universe's density and curvature.

Imagine a triangle drawn on a sphere, like our Earth. The surface curves positively, causing the angles of the triangle to add up to more than 180 degrees. Conversely, a negatively curved universe, resembling a saddle, would have angles totaling less than 180 degrees. Finally, a flat universe, where the familiar rules of Euclidean geometry hold true, would have angles that precisely sum to 180 degrees.

By meticulously analyzing galaxies scattered across the vast cosmic canvas, we were able to estimate the universe's density. We then considered the various curvature possibilities. Surprisingly, data suggests the universe is remarkably flat – at least on a large scale. Despite the warping of spacetime around massive objects like stars and galaxies, when we zoom out to encompass the grand tapestry of galaxies, space appears flat.

Combining this flatness with the influence of dark energy, we arrive at a compelling picture: a spatially flat universe undergoing accelerated expansion. This strongly suggests that the Big Crunch is unlikely. Instead, the universe appears destined for a chilly demise – the Big Freeze – through eternal expansion.

However, the demise of our universe might not be the final curtain. Perhaps from this desolate void, a new universe could miraculously emerge, born from another quantum fluctuation. Even if our universe is unique, a one-time cosmic show, humanity has a vast expanse of time to learn and explore before the final chapter unfolds. Let us continue to refine our telescopes, pushing the boundaries of our knowledge and unraveling the mysteries of the cosmos. The awe-inspiring grandeur that surrounds us is a testament to the universe's vast story, a story that continues to unfold even as we contemplate its potential endings. As we peer into the abyss, the universe stares back, challenging us to decipher its secrets and understand our place within its grand narrative.

But the story doesn't end there. Our curiosity compels us to ask not just about the how and the what, but also the why. Is there a purpose to this grand cosmic dance? Are we, as intelligent life forms, merely bystanders, or do we play a more significant role? Perhaps the answers lie not just in the vastness of space, but also within the depths of our own consciousness. As we peer outwards, we are simultaneously peering inwards, seeking to understand the fundamental nature of reality and our place within it. The universe may be vast and impersonal, yet the human spirit yearns to connect with something greater than itself. We seek meaning in the grand narrative, a purpose woven into the fabric of existence. Perhaps that purpose is the very act of seeking itself, the insatiable curiosity that drives us to explore, to understand, and to create. The universe may be indifferent to our existence, but that doesn't diminish the significance of our journey. In the face of the vast and the unknown, we create our own light, our own stories, our own civilizations. We leave our mark on the cosmos, not through brute force, but through the power of our ideas, our art, and our enduring spirit. The universe may have an end, but the human desire to understand and connect transcends the limitations of time and space. And that, in itself, is a story worth telling.

"As an astronomer and a writer. I have shared all that I can. As per estimates, if you ever want everything into one book, it would be about 1.3×10^6 pages. It would be impossible to write it. Now that you have completed this book, you can say that you are a beginning astronomer. Astronomy, cosmology and Astrophysics are very interesting, fascinating and vast. Everything we ever know, everything we ever see is just 5% of the total universe. So, forget about your problems and live your life without any sorrow, because, all that we do, all that we make, all that we love and all that we leave, matters only on our planet mother earth..."

CREDITS

WRITTEN BY: SREE UDHYAN RAAM

DRAFTED BY: SREE UDHYAN RAAM

ILLUSTRATIONS: SREE UDHYAN RAAM

DEDICATED TO: M.SEETHARAM
           M.SWAPNA
           M.SREE UDBHAV

A.P.J.ABDUL KALAM
   Former president of INDIA, SCIENTIST
NAMBI NARAYAN
   INDIAN SCIENTIST

IMAGES, PICTURES AND SOURCES: INTERNET

COVER INLLUSTRATION: SREE UDHYAN RAAM

COVER DESIGN: SREE UDHYAN RAAM

"This book was just a dedication and contribution to people. Respecting people who have served their best in Cosmology, Astronomy and Astrophysics. Thanking everyone for support and encouragement"

"This is Sree Udhyan Raam Maddineni, signing off..."

www.ingramcontent.com/pod-product-compliance
Lightning Source LLC
Chambersburg PA
CBHW020627220526
45464CB00001B/55